SCIENCE, TECHNOLOGY AND THE MILITARY 1

SOCIOLOGY OF THE SCIENCES

A YEARBOOK

VOLUME XII/1 — 1988

SCIENCE, TECHNOLOGY AND THE MILITARY

Edited by

EVERETT MENDELSOHN

Harvard University

MERRITT ROE SMITH

Massachusetts Institute of Technology

and

PETER WEINGART

University of Bielefeld

KLUWER ACADEMIC PUBLISHERS

DORDRECHT / BOSTON / LONDON

Library of Congress Cataloging-in-Publication Data

Science, technology, and the military / edited by Everett Mendelsohn,
Merritt Roe Smith, Peter Weingart.
p. cm. -- (Sociology of the sciences ; v. 12)
Includes index.
ISBN 9027727805 (v. 1) ISBN 902772783X (v. 2)
1. Military art and science--History--20th century.
2. Technology--History--20th century. 3. Technology and state-
-History--20th century. 4. Munitions--History--20th century.
5. World politics--20th century. I. Mendelsohn, Everett.
II. Smith, Merritt Roe, 1940- . III. Weingart, Peter.
IV. Series.
U42.S35 1989
355'.009'04--dc19 88-12876

ISBN 978-90-481-8454-5

Published by Kluwer Academic Publishers,
P.O. Box 17, 3300 AA Dordrecht, The Netherlands

Kluwer Academic Publishers incorporates
the publishing programmes of D. Reidel, Martinus Nijhoff,
Dr W. Junk, and MTP Press.

Sold and distributed in the U.S.A. and Canada
by Kluwer Academic Publishers,
101 Philip Drive, Norwell, MA 02061, U.S.A.

In all other countries, sold and distributed
by Kluwer Academic Publishers Group,
P.O. Box 322, 3300 AH Dordrecht, The Netherlands

printed on acid free paper

TABLE OF CONTENTS

VOLUME XII/1

VOLUME XII/2

PART III

Transformation of Industry and Medicine

PART IV

Nuclear Weapons and Nuclear Power

PART V

R&D: Military, Industry and the Academy

PREFACE

The extensive interaction between science/technology and the military has become increasingly apparent in the years since the second World War. New institutional arrangements, new fields of study and research, new patterns of funding and support, new relations with industry, the academy and the state, and new professional roles have marked the sciences and technology; the military transformations have been equally important: new weapons systems of great complexity, new strategies and practices, new reliance on high technologies and advanced sciences, and extensive involvement in the funding of science and technology in the "civilian" sector. While the literature in the social studies of science and technology has from time to time addressed these issues (especially from the historical perspective), the attention paid to the very extensive interactions and concommitant transformations has not been commensurate with the magnitude of the enterprise or changes undergone. Our present volume has the intent of both contributing reports of new research and stimulating additional study.

Two recent volumes deserve explicit notice since they represent significant contributions and servé as models for future work. At the XVII International Congress of the History of Science, Berkeley, California, August 1985, a number of papers focusing on the science/technology-military relationship were presented in a symposium on Cooperative Research in Government and Industry. A group of these papers has been brought together and published in *Historical Studies in the Physical and Biological Sciences*, Volume 18, Part 1, 1987, edited by J. L. Heilbron. Another collection, edited by one of us, provides both a series of detailed cases and a valuable bibliographic essay: Merritt Roe Smith (ed.), *Military Enterprise and Technological Change* (Cambridge, Mass.: M.I.T. Press, 1985).

The papers published below were all presented at a conference organized by the *Yearbook, Sociology of the Sciences*, and held in

Cambridge, Mass., 8—10 January 1987, in the quarters of the Science, Technology and Science Program at M.I.T. We are grateful to that program and to the Department of the History of Science at Harvard University for their helpfulness and cooperation in planning, organizing, and holding the conference.

The workshop was supported by a grant from the John D. and Catherine T. MacArthur Foundation, for which we are deeply grateful. It reflects the Foundation's broad concept of security studies. This volume was very helpfully edited by Julia McVaugh. Much of the secretarial staff work behind the conference and the publication has been thoughtfully done by Ruth Bartholomew.

EVERETT MENDELSOHN
MERRITT ROE SMITH
PETER WEINGART

INTRODUCTORY ESSAY
SCIENCE AND THE MILITARY: SETTING THE PROBLEM

EVERETT MENDELSOHN

Harvard University

MERRITT ROE SMITH

Massachusetts Institute of Technology

and

PETER WEINGART

University of Bielefeld

I. Science for War?

In the closing days of his presidency, Dwight Eisenhower (a World War II military hero) warned his fellow American citizens of two interlinked dangers: the emergence of a military-industrial complex that would have undue influence on national policy, and the development of a scientific-technological elite whose knowledge of advanced military systems would usurp the role of civilian leadership in critical areas of decision making. While scoffed at by many at the time, the predictions have by now achieved the status of a truism.

Even as President Reagan's recent plan for a Strategic Defense Initiative (SDI, or commonly known as "Star Wars"), which would render all nuclear weapons superfluous, made the news headlines, many were able to recognize signs once again that the arms race was propelled not so much by the perception of "threat" from the Soviet Union, as by the ambitions and enthusiasms of scientists and engineers. Since World War II, there have been many complaints by secretaries of defense that weapons systems had become so complicated and so

E. Mendelsohn, M. R. Smith and P. Weingart (eds.), Science, Technology and the Military, Volume XII, 1988, xi—xxix.

expensive that the smaller (NATO) countries could not afford them any more. There were further doubts about the serviceability of high-technology weapons in the event of war and warnings that arms-control negotiations were being hampered by continuing rapid new developments in the arsenals. To observers and analysts of science and technology, these signs fell in place alongside other evidence going back to the origins of the present research and development complex.

The emergence of systematic national science policies during the 1950s was closely linked to the transfer of wartime R&D capacities to civilian uses — both in terms of knowledge and technology, as in the case of nuclear power, and in terms of institutions, such as large government and industrial laboratories. For the first time in history, the entire span of activity from supposedly "basic" scientific research to weapons production had been institutionalized in systematic fashion on an industrial scale. This legacy was preserved politically with the conviction that any future war would be won by the power that commanded the best scientific and technological resources — that is, scientific knowledge, technical know-how, and productive capacity. The announcement of the "Star Wars" scheme by President Reagan was but a reminder that this arrangement was still fully in place and that at least some scientists were selling technological fantasies to politicians and generals. If the fantasies were accepted, the scientists would gain financial resources as well as long-term, large-scale projects on which to work — projects of great complexity, challenge, and thus attractiveness. When President Reagan accepted "Star Wars," he tried to sell it to the public as the ultimate weapon, which in the end he would even share with the enemy. But those who remained skeptical of the recurring promises of "ultimate weapons" saw in SDI another example of science and technology driving the arms race and preventing disarmament accords (1).

While this view holds little surprise for some, it should be remembered that it runs counter to common belief. Political ideology is in many ways ideally complementary to the ideology of science in that the former makes us believe that all military preparations are a direct response to the military threat from outside, with science merely serving to develop needed tools. As far as science and technology are involved in weapons making, they see themselves only in a service role in the national interest, performing a patriotic duty. The ideology of science goes one step further and points to the fact that while science may be utilized for military ends and may thus be involved in destruction, its ultimate purpose is still the pursuit of truth for the benefit of mankind — an essentially humanistic endeavor.

The perspective of science and technology assuming the role of the propelling force of the arms race was the initial rationale for undertaking our effort to direct the attention of scholars in the "social studies of science" to the relation between "science, technology, and the military." While historical studies of this relationship abound; while political analysts have examined in detail the political conditions of the integration of science and technology into the war effort; and while analyses of the development of strategic thinking in the missile age fill entire libraries, few scholarly efforts bring to bear the advances in the "social studies of science" on understanding the new relationship between science and the military. No full studies put science and technology in the center of analysis and look at their role in shaping the thinking and action of military and political leaders.

Given this perspective, this volume is a beginning. But even a superficial involvement with the problem reveals that the initial perspective is too simple, as indeed almost any monocausal model would be. Not surprisingly, calling on those who work on some aspect of the science-military connection forced the recognition of different kinds of evidence. Furthermore, since our endeavor is limited and did not have recourse to a large-scale research effort, we make no claim to offer a representative cross section of "work done in the field" or to provide a definitive statement of the "state of the art." Rather, this is a first attempt to bring together a group working on the issue and to give further encouragement and focus to research on the topic of "science, technology, and the military."

II. Patterns of the "Science-Military" Interaction

Any discussion of different cases and comparative historical examples reveals that the relationship between science and the military is in constant flux, and the very natures of these two institutional sectors of society themselves undergo change. There are, in addition, cultural and political contexts to contend with.

Comparative historical studies are called for, to focus on the respective roles of science, the military, and industry in times of peace and war: before World War I, during the interwar years, in World War II, and in the present arms race. By observing the changes of the respective boundaries, shifts in relative weight and in the nature of operation become apparent. There can be no doubt that the military has become more "scientific" (and technological), but likewise science has become significantly militarized, at least in parts. Thus, the nature of both institutions has changed profoundly because of their reciprocal relationship, and the problem of the political system needed to control the military has changed as well. Similarly, it is often said that science has become industrialized (both in terms of sheer size of the research establishment and in terms of the work organization of research), and conversely we speak today of "science-based industries," denoting the scientification of industry in many of today's key areas of production. Much of this process can be traced to the joint involvement of science and industry in the war effort (2).

How, then, can we account for what appears to be a growing interpenetration of science and the military? What, if any, are the patterns of this process? What are the consequences for science, the military, industry, and the political system?

First of all, speaking about "science" and the "military" presupposes an institutional identity that delineates their boundaries — and this is a problem precisely because the primary interest of analysis is the changes that occur between them. On the one hand, it is apparent that by looking at the institutional connections and interrelations their very boundaries appear fuzzier the more closely we look, or the more extended a historical view we take. On the other hand, the institutional divisions are real, they have emerged over time, and they stand for important differentiations of purpose, values, and activities. Thus, one has to know *what* changes are being looked at, and in what respect they are being examined. The theory of social systems — which is the most advanced and powerful tool of the social sciences, and in its more recent development allows us to bridge the gap between so-called "micro-" and "macro"-analysis — can give useful guidance in this instance. Rather than attempting to provide ad hoc definitions, which lead into the problems mentioned, it is advantageous to proceed from a more abstract level of analysis. Science as a social system has become institutionalized on the basis of the binary schematism of "true-untrue" — that is, the socially established criterion of truth versus nontruth. This by no means presupposes an essentialist definition of truth, which does not exist; rather, it simply means that people belonging to that system act on the assumption that it is possible to differentiate between truth and, say, ideology or power. That assumption structures their activities, their communication, and their institutional arrangements. Likewise, the military, which must be seen as an extension of the political system, is based on the schema of power-nonpower, and the same considerations concerning its organization and operation must be applied to it. Finally, if we want to include industry as the third institutional sector involved in the setting we are studying, it is based on the schema of economical-noneconomical; ideally neither truth nor political (or military) power is of concern to the operation of industry.

The problems arising from the interaction between the three systems all have to do with the consequences of the fundamental incompatibility of these schemata and are readily apparent: an intrusion of the military into science may lead to demands for secrecy, thereby disrupting an efficient scientific communication through the danger of the suppression of "true" knowledge or, even worse, functioning technological solutions; an undue dependence of industry on military contracts may lead to its failure to compete on the market because of not being able to meet cost-conscious standards of production; a strong reliance of the military on scientific and technological advances may create a weapons arsenal that, although extremely sophisticated, will not be serviceable in battle or flexible enough to adapt to changes in strategic concept. These are the kinds of problems arising at the "interface" of the systems when the boundaries are shifted in one or the other direction.

But just as the institutional identities of science, the military, and industry have emerged historically and are not fixed, they are continuing to evolve. The differentiation between them, as we know it now, is surely not the endpoint of social development, and it is quite possible that the interrelations between them in which we are interested signal a new era. The reciprocal expansion of the systems under consideration is a common characteristic of such systems, once they have achieved sufficient independence. In search of stability, they attempt to control their environment by bringing additional areas within their boundaries.

It goes without saying that it is a particularly relevant environment that is the target. Any conflict over boundaries is preceded by some realization of the utility of the particular environment for the system under consideration. When Fritz Haber offered his services as a chemist to the German army during World War I, pointing to the potential uses of poison gas, it took some time before the military staff adopted the argument because it presupposed a change in important ways of thinking. Once the utility of poison gas was recognized among military ranks, the alliance with the scientists was made (3). The process can be outlined: scientific research groups offer scientific/technical solutions to military problems — often redefining them — in order to obtain research and development grants, and thereby to expand their research installations and attract larger numbers of students; the military, in turn, tries to gain control over those research groups which offer interesting prospects for military objectives, both by funding them (or even creating in-house research laboratories) and by exerting bureaucratic controls, such as security clearances; industrial companies, for their part, try to gain control over the military by obtaining monopolies on specific technologies in order to win long-term procurement contracts that make them relatively independent of the market; and so forth. These inherent tendencies to extend their boundaries create a constant process of tension and conflict that virtually characterizes the relationships between science, industry, and the military, most notably since these relationships became a systematic element in the post-World War II period (4).

The logic of this analytical framework poses the next question: By which means does each system manage to extend its boundaries? While some systematic answers can be given, it is here that historical analysis can provide evidence for the model or counterevidence against it. Comparative historical analysis alone is able to produce empirical evidence regarding the changes of the military, of science, and of industry within themselves and relative to each other. It thus gives the background against which one can look for the particular mechanisms effectuating the shifts among them.

Virtually any expansion of one system "at the cost of another" depends on the coincidence of their dynamics and constellations in the environment that supports them. Some aspects of the "scientification" of the military clearly result from the fact that the military faces problems that science has been able, or at least has claimed to be able, to solve. The obvious and perennial problem of keeping up troop morale, which was traditionally dealt with by using moral appeals (if not drastic sanctions), can be treated – though perhaps not entirely solved – by turning to therapeutic psychiatric forms (cf. Edward M. Brown's paper in this volume). By replacing a moral vocabulary with a scientific one, the functional problem of military discipline is transferred to science and thus "objectified." This process undoubtedly relieved the military of the burden of justifying traditional means of discipline in a world in which such justification was becoming increasingly difficult (5).

A current example of establishing links to the environment on the part of the military is provided by the case of space technology. This technology, in particular, lends itself equally well to civilian and military purposes. In the case of both SDI and the European space program it can be assumed that the community of space scientists and engineers and the military tie their respective goals to those of the other party, thus gaining public support at a level that would not otherwise be available (cf. Johannes Weyer's paper in this volume). In short, where there is a high political legitimacy for warding off a military threat in space, the scientific and technological project is justified in terms of its military uses; and where the political priority is on civilian space technology, the project is justified with civilian uses. In each case, the duality of function is played down. The most dramatic instance of this kind is, of course, war itself; when war is imminent or already under way, and no longer just an assumed threat, it allows the political system to mobilize enormous resources at the cost of almost any other social priority (6).

Thus, war greatly enhances the legitimacy of military demands and contributes to the expansion of this legitimacy. The "militarization" of physics in the wake of the construction of atomic and nuclear bombs has been thoroughly analyzed, and the case has become a historical paradigm of the shift in the constitution of science as well as of science policy (7). Several aspects of the story are of interest for our problem, one of them undoubtedly being the ability of the military to extend its control far into the realm of science, supported by the scientists' willingness to comply. A detailed historical and biographic look at the motivations and activities of protagonists like James B. Conant and J. Robert Oppenheimer reveals the repercussions of this process at the person level (cf. James Hershberg's paper in this volume) (8).

In order to specify further the mechanisms mentioned, it helps to abstract from concrete examples. We can differentiate between two types of considerations that can be brought to bear in the quest for expansion: (1) legitimation, and (2) knowledge. They are not mutually exclusive; on the contrary, they may be combined. It is precisely their interrelation that ties the scientific and military systems to their respective environments and accounts for different constellations.

(1) The military relies heavily on the legitimation gained from postulating the threat of annihilation. While this is indisputable in times of war, it carries less force in peacetime, especially as the price tag for military preparedness is rising constantly. The threat has to be "constructed" by pointing to the preparation of the putative enemy. Thus, the logic of the arms race is the mutuality of interests of the military establishments on both sides, who supply each other with the legitimacy for their claims.

The particular legitimation of military preparedness may be carried over into science and become part of its expansionist strategies, especially in times of peace, when few other sources of funding are available — that is, when the legitimacy of claims to substantial public support for basic research is low. This appears to have been the case after the Second World War in the United States, and it led to the formation of a strong alliance between U.S. science and the military; the pressure on science was all the more acute because the science system had expanded significantly during the war, and questions arose concerning the fate of the big research laboratories and the prospect of wasting the existing capacities for training a whole generation of scientific manpower, as well as losing the services of those already educated. The Office of Naval Research (ONR) in large measure reflects the new alliance. An interesting aspect of this particular story is the fact that the transfer of the legitimacy of military preparedness into science also changed a number of legitimating values of science proper: involvement in military research became legitimate for scientists *even in peace time*. Not until the Vietnam War was that legitimating pattern challenged, and despite that challenge it has left the military with a substantial part of the scientific establishment serving its needs (9).

This latter situation points to a split within the science system with respect to the legitimating values used. Although the legitimation of public support for research and training has greatly increased, it has done so with reference more to the link with technological development and economic well-being — that is, civilian uses — than to "knowledge for its own sake." Therefore this type of legitimation is not exclusive to science as a social enterprise, but connects it with other systems, which in turn may be a source of vulnerability for science. This was certainly the case during the Vietnam War when the reigning national political consensus was broken.

Thus, both current and historical analyses of the interrelation of science and the military must focus on the processes by which the military takes control of or penetrates the scientific system as a result either of contingent developments or of intentional "managing," as well as on how science "hooks on" to the political legitimation of the military in order to advance its own purposes.

(2) Knowledge as a resource would seem to be primarily a function of science and to provide it in turn with the means to penetrate other systems. The power of science to provide new technical options or to demonstrate solutions superior to those traditionally used puts it in a position potentially to gain the power to actually define problems. There are surely many instances, as in the case of Project Hartwell, where military programs are triggered by civilian scientists — that is, where a knowledge-based technological option emerges from the science system and forces adaptation on the part of military thinking.

This is the kind of dynamic that contributes to the impression that an important part of the arms race is contingent upon scientific-technological development rather than upon military and political strategies. A notorious case in point is the "Chevaline" project in Great Britain, which led Lord Solly Zuckerman to his claim that the ideas for new weapons systems emanate from groups of scientists and technologists in government labs and industry, and not from the military (see Philip Gummett's paper in this volume) (10).

Zuckerman's statement highlights the analogous situation with respect to the relationship between industry and the military, as seen especially in the case of the U.S. contract-based system of weapons acquisition, where the technological know-how and traditions of particular firms may strongly influence the military's perception of new armaments and their strategic uses (cf. Judith Reppy's example of Teledyne-Ryan's RPVs and the Air Force) (11).

"Scientification" is characteristic not just of the military, but of modern societies in general. The increase in importance of knowledge as a resource is neither a new insight nor terribly surprising. What *is* surprising, however, is that very little is known publicly about the role of science and scientists in relation to the military. Of course, the ideology of science traditionally does not allow science to be perceived as too deeply involved with the military system. This may explain why an almost artificial conflict is seen between explanatory models that build on the two types of resources. Fashioned very much along the lines of the classical "market pull" and "technology push" dichotomy, the debate is over the issue of whether the military leads the technological weapons development via its legitimatory resources of "threats," or whether the process is actually "pushed" by developments "internal" to science and technology (cf. Gummett's paper, which contains both theses). Clearly, the analysis of the science-military connection desperately needs case studies (which may support both models), both to establish and to broaden the empirical base on which further theorizing can be built. But in the longer run it may well be more fruitful to look for models that transcend thinking in terms of dichotomies.

If one takes the evidence at hand, it becomes apparent that the development of the science-military-industry connection (not to say the scientific-military-industrial complex) owes its emergence and spectacular growth to mutual reinforcements. First of all, it is misleading to view the two principal resources, legitimation and knowledge, as exclusive of each other. On the contrary, in the context of highly developed capitalism and the two-superpower constellation, they reinforce each other. The philosophy that emerged at the end of World War II holding that military preparedness is tantamount to scientific and technological superiority has gained even greater strength during the Reagan era. In essence it means that the war of the future will be decided not on the battlefield, but in the research labs. By this logic, science has been drawn into the legitimating framework of militarization indirectly, but so irresistibly that it is hard to imagine how the status quo ante can ever again be restored.

Likewise, on the level of knowledge it is important to realize that there is no longer, if there ever was, an ontological distinction between purely scientific knowledge and militarily relevant knowledge (or just "applied" knowledge and technology). The cases of "Star Wars" technology, and of space science and technology, only repeat the historical experience of the atomic bomb – namely, that it matters little whether a certain technology is developed for military or for civilian purposes; in each case, "pay-offs" and "fall-outs" may go either way. The fact that today's high technology is no longer clearly distinguishable from science, and that distinctions between them are refuted on the level of patents and scientific literature, suggests that the dynamics of economic development and weapons development have their source in a general stock of scientific/technical knowledge that may be utilized and expanded for a multitude of functions and purposes, both civilian and military (cf. Peter Galison's paper in this volume) (12).

This view is supported on the institutional level as well. The co-operation between science and the military begun during the Second World War created a pattern of circulation of elites, at least in the United States, that draws the envy of generals, senior scientists, and industrial leaders in other countries: many attempts, some successful, have been made at copying the model. The coincidence of scientific/technical potentials offered by scientists and the ability of farsighted political and military leaders to take up this opportunity has led to the expansion of the science system and has made permanent this circulation of elites (cf. Silvan Schweber's paper) (13). A closer study of these elites would almost certainly reveal differences in style, goal orientation, and interests. But when one tries to account for the close connections between science, the military, and industry, the most interesting elements are the commonalities across the boundaries of the respective social systems to which they belong, which enable at least a certain stratum among them to move across these boundaries quite freely. These commonalities are based on common interests that tie their career fates together.

Certainly, a specific industry may see the dangers of being exclusively dependent on defense contracts and of losing the capability to compete in the civilian market; but experience seems to show that this fear disappears very rapidly in times of economic recession. Science establishments outside of the government — in universities, in particular — may be reluctant to rely entirely on military research contracts, but, like industry, they have expanded due to military contracts to such an extent that to live without them would imply severe contraction (14).

Governments, of course, do their best to keep these dependencies effective. A program like Independent Research and Development (IR&D) is designed to do just that, and it is in itself a proof of the fact that it is enough to keep an R&D capacity alive and well, without trying to steer it rigidly to preconceived tasks, in order to preserve the knowledge base deemed to be crucial for military preparedness. (It is ironic that through this program the military actually follows the legitimation pattern once advanced by science: namely, that the best way to get pay-off from research is to leave it alone.)

Another important commonality is the similarity, if not identity, of the professionalism that rules the career perspectives in all three spheres, and which is a sign of the knowledge base of their respective realms of activity. Professionalism as an organizing principle of patterns of action permeates modern societies and accounts for common outlooks, social bonds, and mutual understandings. Taken together with institutional arrangements for research and engineering that are similar in structure and function both inside the government and outside, it may be surmised that the similarities in orientation, and the overlaps of interest, are sufficient to guarantee the "multifunctionality" of managerial and scientific/technical manpower.

III. The Science-Military-Industry Connection as a Field of Study

Our remarks are almost entirely based on a small amount of unsystematic evidence, informed guessing, and theoretical extrapolation. The explicit and implicit questions point to the desperate need for sociological and historical, comparative research focusing on the relationships among science, technology, the military, and industry.

Part of the explanation for the fact that little attention has been given to this complex is the preoccupation with academic scientific research and the more or less unquestioned assumption that such research has no connections with technology. Even when that assumption came up for debate, a long and fruitless discussion evolved over the independence or autonomy of one field from the other and over their essential differences. Most of that debate was probably motivated by the respective interests and limited horizons of historians of science and historians of technology. The realization that in terms of both cognitive and institutional developments the conceptual boundaries are beginning to be blurred, is fairly recent (15).

Another reason for the paucity of research by historians and sociologists of science and technology is the general acceptance of the ideology that science is a fundamentally humane endeavor, and that its involvement in war is, at best, a patriotic necessity limited in duration of time, and at worst, an aberration. Only the Vietnam War disturbed this belief. Studies of the contributions of scientists and engineers to military activities, as well as studies of their profiting from wartime conditions and connections that are hard to justify in terms of patriotism, are just now emerging (see, for example, the collaboration between science and the military in Germany under the Nazis) (16).

Now that both obstacles seem to be overcome, and with the relevance and urgency of the science-military issue being underscored by yet another acceleration of the arms race, the field is wide open for more systematic analysis. Some problems that seem particularly important can be enumerated:

Attention should be given to the processes from which new military technologies emerge. As Donald MacKenzie's case study (in this volume) of the development of the "Ballistic-Missile Stellar-Inertial Guidance System," and to some extent I. B. Cohen's account of the relation between computer scientists and the military, are revealing, it would be extremely important to elucidate these processes in order to improve our understanding of the interaction among scientific, civilian, and military objectives in the shaping of new technologies. Insight into these processes of negotiation will be relevant both to theoretically oriented research on the "social construction" of science and technology, and to understanding the role of science in weapons development.

Another extremely important issue is the impact of an orientation to military needs on science and technology. As Leon Trilling shows (in this volume), the perception of economic, geographic, and manpower training conditions and the definition of military purposes shaped the specific design of technologies in fighter aircraft. The same can be said of civilian technologies. From this evidence it can be deduced that, given an "underdetermination" of technical solutions — that is, their principal variability — an orientation to coherent frameworks of functions, such as military ones, will have a penetrating impact on the scientific and technical problems that are selected, on the ways in which they are conceptualized, and on the solutions that are deemed satisfactory. This entire complex of the cognitive impact of a military "mind-set," which previously has been seen, if at all, only in terms of the "spin-off"-oriented studies that were supposed to legitimate expenditures for military research, is as important as it is difficult to analyze (17).

The notion that both industry and the military are drawing on a general stock of scientific and engineering knowledge, which Galison advances in his study of the bubble chamber in high-energy physics, is worth exploring in greater depth. It could be interpreted to presuppose an autonomous development of technology, which would be contradictory to the aforementioned approach of the "social construction" of technologies. Consequently it would be important to put this notion to a test — for example, by identifying the institutional locus where this stock of knowledge originates and by showing the communication flows between the different sectors. Studies of this kind should be expanded so as to determine the "multifunctionality" and availability of particular stocks of knowledge as crucial inputs into the arms development process.

The same issue may be analyzed on the level of the institutional mechanisms that are designed to facilitate, or that hamper, the "drawing on a stock of knowledge." Here, in particular, comparative and historical studies are called for, as the phenomenon seems to be historically fairly recent, and different political systems seem to operate in very different fashions. The American system of securing scientific and technical capabilities for the military obviously differs from the British and German systems; it may be asked, can one observe processes of adaptation of the latter to the former due to demonstrated success? By comparing different institutional arrangements of this sort we might be able to determine whether we are witnessing a move from a system of government-laboratory-based defense R&D to one military funding of independent R&D; that is, from the support of an "in-house" capability to the tapping of the national science and engineering capacity as a whole. Of course, these are not mutually exclusive, but significant shifts in support will point to shifts in underlying "philosophy."

This is but one example of the way in which the analysis of funding patterns can serve as an important starting point for much deeper interpretation than at first seems apparent. In fact, the first attempts to look at the science-military connection usually began with analyses of defense-related R&D budgets. Needless to say, it is informative in itself to gain an understanding of the financial magnitudes involved, as Gummett provides it on a comparative basis for the United States and Great Britain. That is just the base line, however. If the budgets are analyzed on a "micro" level — for example, in connection with the types of funding mechanisms, with the organizational structures of R&D, and with the types of programs that are funded — conclusions may be drawn in a systematic fashion about which type of knowledge is demanded by the military from which types of institutional sources, and which type of manpower is involved in delivering this knowledge.

Like the analysis of institutional structures and their interconnection, that of the "elites" promises to be fruitful for a deeper understanding of the relationships between science, industry, and the military. The problem is not primarily that of traditional "elite studies," but rather in what respects they differ and what their common interests and outlooks are. If the wartime collaboration (in the case of the United States), as shown in Schweber's historical study, can be generalized, the significance of analyses of the elites is to show the similarity of scientific and engineering objectives in the different sectors, and the ways in which stocks of knowledge are communicated and functionalized for seemingly very different purposes. "Elite circulation," to use Vilfredo Pareto's terminology, is not interesting primarily as revealing a "conspiracy"-like rule of technocrats, although that revelation may be interesting in its own right, but rather because it is indicative of a common knowledge-base linking the scientific, military, and industrial systems.

These are some of the research problems that seems to us to be in need of analysis. Undoubtedly, as scholars turn to the science-military-industry connection, they will discover many other problems that will prove to be equally interesting. It is hoped that the information barrier surrounding this complex can be at least partially overcome, not least by the ingenuity of those willing to risk taking on the task. The ultimate objective of such an endeavor cannot be to advance knowledge per se, as important as that may be, but rather to reveal in scholarly fashion, and with results that stand up to critical analysis, the intricate involvement of science with the "weapons culture." In this manner science studies may help to increase the political and ethical awareness of those who are part of that "culture," and to make the choice of membership in it a clearer one for them.

Notes to Introduction

1. A huge literature exists on "Star Wars." See, for example, William J. Broad, *Star Warriors* (New York: Simon & Schuster, 1985); Katie Schwarz, "Panelists Discuss Economics of SDI," *The Tech* **105** (Nov. 22, 1985): 1, 17; Peter H. Stone, "'Star Wars' Draws New Fire," *Boston Globe*, January 4, 1987; Douglas Waller et al., "SDI: Progress and Challenges: Staff Report Submitted to Senator William Prox-

mire, Senator J. Bennett Johnston, and Senator Lawton Chiles" (Unpublished paper, U. S. Senate, March 17, 1986); Fred Kaplan, "'Star Wars': The Ultimate Military-Industrial Compact," *Boston Globe*, September 14, 1987.

2. For entry to the literature on science-based industries, see David F. Noble, *American by Design: Science, Technology, and the Rise of Corporate America* (New York: Oxford Univ. Press, 1977); George Wise, *Willis R. Whitney, General Electric, and the Origins of U.S. Industrial Search* (New York: Columbia Univ. Press, 1985); Leonard S. Reich, *The Making of American Industrial Research: Science and Business at G.E. and Bell, 1876—1926* (Cambridge: Cambridge Univ. Press, 1985); Stuart W. Leslie, *Boss Kettering* (New York: Columbia Univ. Press, 1983). On the involvement of science and industry in war, see William H. McNeill, *The Pursuit of Power* (Chicago, 1982); and Merritt Roe Smith (ed.), *Military Enterprise of Technological Change: Perspectives on the American Experience* (Cambridge, Mass.: M.I.T. Press, 1985), especially chapters 4, 6, and 8. Also see the papers prepared for the NATO Advanced Workshop on the Relationship Between Defense and Civil Technologies, Wiston House, Sussex, Sept. 21—25, 1987.
3. See L. F. Haber, *The Poisonous Cloud: Chemical Warfare in the First World War* (Oxford: Clarendon Press, 1986) and Gilbert F. Whittemore, Jr., "World War I, Poison Gas Research and the Ideals of American Chemists," *Social Studies of Science* **5** (1975), 135—163. Also see Daniel P. Jones, "From Military to Civilian Technology: The Introduction of Tear Gas for Civil Riot Control," *Technology and Culture* **19** (1978): 151—168.
4. See the collections of papers in: Monte D. Wright and Lawrence J. Paszek (eds.), *Science, Technology, and Warfare* (Washington, D.C.: Office of Air Force History, 1971); Franklin A. Long and Judith Reppy (eds.), *The Genesis of New Weapons: Decision Making for Military R&D* (New York: Pergamon, 1980); John Tirman (ed.), *The Militarization of High Technology* (Cambridge: Ballinger, 1984). For other examples, see Thomas Misa, "Military Needs, Commercial Realities, and the Development of the Transistor, 1948—1958," in M. R. Smith, *op. cit.*, 1985 (2), 253—287; Michael A. Dennis, "No Fixed Position: University Laboratories and Military Patronage at Johns Hopkins and MIT, 1944—1946" (Paper presented at the annual meeting of SHOT, Raleigh, N.C., Oct. 31, 1987).
5. See Peter Buck, "Adjusting to Military Life: The Social Sciences Go to War," in M. R. Smith, *op. cit.*, 1985 (2), 203—252.
6. See, for example, Michael Smith, "Selling the Moon: The U.S. Manned Space Program and the Triumph of Commodity Scientism," in Richard W. Fox and T. J. Jackson Lears (eds.), *The Culture of Consumption* (New York: Pantheon, 1983), 177—209; Walter A. McDougall, *The Heavens and the Earth: A Political History of the Space Age* (New York: Basic Books, 1985).
7. See Daniel Kevles, *The Physicists: The History of a Scientific Community in Modern America* (New York: Knopf, 1978); Spencer Weart, *Scientists in Power* (Cambridge, Mass.: Harvard Univ. Press, 1979); Martin Sherwin, *A World Destroyed: The Atomic Bomb and the Grand Alliance* (New York: Knopf, 1975).
8. James Hershberg, in addition to the paper in this volume, has published several other important studies focused on Conant, "James B. Conant and the Atomic Bomb," *Journal of Strategic Studies* **8** (March 1985), 78—92. See also James B.

Conant, *My Several Lives: Memoirs of a Social Inventor* (New York: Harper & Row, 1970); Peter Goodchild, *Robert J. Oppenheimer: Shatterer of Worlds* (Boston: Houghton Mifflin, 1981); Herbert York, *The Advisors: Oppenheimer, Teller and the Super-Bomb* (San Francisco: Freeman, 1976).

9. See James Phinney Baxter, *Scientists Against Time* (Boston: Little, Brown, 1946); Edward Salkovitz (ed.) *Science, Technology and the Navy: Thirtieth Anniversary 1946—1976* (Arlington, VA: Dept. of the Navy, Office of Naval Research, 1976); Guy Hartcup, *The Challenge of War: Britain's Scientific and Engineering Contribution to World War II* (New York, 1970); Ralph E. Lapp, *The New Priesthood, the Scientific Elite and the Uses of Power* (New York: Harper & Row, 1965); Ralph E. Lapp, *The Weapons Culture* (New York, 1968); H. L. Nieburg, *In the Name of Science* (Chicago: Quadrangle, 1966); for recent critical views, see Fred Kaplan, *The Wizards of Armageddon* (New York: Simon & Schuster, 1983) and Grey Herken, *Counsels of War* (New York: Knopf, 1985); see also Herbert F. York, *Making Weapons, Talking Peace: A Physicist's Odyssey from Hiroshima to Geneva* (New York: Basic Books, 1987). For an introduction to the historical literature on the United States see: Alex Roland, "Science and War," *Osiris 1*, Second Series, 1985, pp. 247—272.
10. Solly Zuckerman, *Nuclear Illusion and Reality* (London: Collins, 1982), p. 103.
11. Samuel L. Hall, "Weapons Choices and Advanced Technology: The RPV," Peace Studies Program Occasional Paper No. 10 (Ithaca, N.Y.: Cornell University Peace Studies Program, 1978). See the additional studies in Franklin A. Long and Judith Reppy *op. cit.*, 1980 and Kosta Tsipis and Penny Janeway (eds.), *Review of U.S. Military Research and Development, 1984* (Washington, D.C.: Pergamon-Brassey's, 1984).
12. See also M. R. Smith, *op. cit.*, 1985 (2), 34—36.
13. In addition, see the Kevles and M. R. Smith volumes mentioned above and also Carroll W. Pursell, Jr. (ed.), *The Military Industrial Complex* (New York: Harper & Row, 1972); Stephen Rosen (ed.), *Testing the Military Industrial Complex* (Lexington, Mass.: D. C. Heath, 1973). The integration of science and the military in Britain is chronicled in Ronald W. Clark, *The Rise of the Boffins* (London: Phoenix House, 1962).
14. See, for example, "Report of the Ad Hoc Committee on the Military Presence at MIT" (from the Office of the Secretary of the Faculty, May 1986); also see Dennis *op. cit.*, 1987 (4).
15. See, for example, John M. Staudenmaier, S. J., *Technology's Storytellers* (Cambridge, Mass.: M.I.T. Press, 1985), especially chapter 3; also Wolfgang Krohn, Edwin T. Layton, Jr. and Peter Weingart (eds.), *The Dynamics of Science and Technology, Sociology of the Sciences Yearbook 1978* (Dordrecht: Reidel, 1978).
16. There is a growing literature in this field. For an introduction, see Alan D. Beyerchen, *Scientists Under Hitler*, Politics and the Physics Community in the Third Reich (New Haven: Yale Univ. Press, 1977); Karl-Heinz Ludwig, *Technik und Ingenieure im Dritten Reich* (Königsten: Athenäum, Dusseldorf: Droste, 1974, 1979); Herbert Mehrtens, "Das 'Dritte Reich' in der Naturwissenschaftsgeschichte: Literatur und Problemskizze," in Herbert Mehrtens and Steffen Richter (eds.) *Naturwissenschaft, Technik und NS-Ideologie*, (Frankfurt: Suhrkamp, 1980) 15—87. Although the Vietnam War years saw the publication of many books and

papers critical of science and technology at both the socio-political and epistemological levels, there is as yet no systematic examination of this "literature of criticism."

17. In addition to Trilling's essay in this volume, see M. R. Smith, *op. cit.*, 1985 (2) for further entry into these problems.

PART I

WAR AND THE RESTRUCTURING OF PHYSICS

THE MUTUAL EMBRACE OF SCIENCE AND THE MILITARY: ONR AND THE GROWTH OF PHYSICS IN THE UNITED STATES AFTER WORLD WAR II

S. S. SCHWEBER

Brandeis University

World War II altered the character of science in a fundamental and irreversible way. The importance and magnitude of the contribution of scientists and engineers — particularly physicists — to the American war effort, changed the relationship between scientists and the military, industry, and government (1). During the war a close relationship developed between the scientific community and the military. After the war ended, the Department of the Navy and the War Department, realizing that the security of the nation depended on the strength and creativity of the scientific community and its institutions, invested heavily in their support and expansion (2). By the mid-fifties a new generation of scientists and engineers had been trained. Many of them staffed the government and industrial laboratories that carried out the research and development for the armed forces. Control of these laboratories and control over the procurement process that supplied their weapons allowed the armed forces to regain control over the planning and deployment of new weapons systems. During World War II they had lost some of this control to the office of Scientific Research and Development (OSRD) and the other civilian boards that had been responsible for much of the innovation in weaponry. The trial of J. R. Oppenheimer made it clear that a new set of rules governed the relationship between scientists and the military.

The partnership between the military and the scientific community had been forged during the war by the men who oversaw the scientific war effort: V. Bush, J. B. Conant, F. B. Jewett, R. C. Tolman, and K. T.

E. Mendelsohn, M. R. Smith and P. Weingart (eds.), Science, Technology and the Military, Volume XII, 1988, 1—45.

Compton (3). This elite shared the views of the then leaders of the armed forces, men such as General G. Marshall and Secretary of War H. Stimson. They all were conservative, and they all took for granted the civilian control of military affairs. They all firmly believed that the armed forces served the nation and were responsible to the president and accountable to the Congress. Furthermore, these men all agreed on the role the United States was to play in maintaining the peace once victory had been achieved.

After World War II the alliance was cemented by the scientists who had been in charge of the laboratories that created and perfected the science-based weapons that had won the war, among them: J. R. Oppenheimer, E. O. Lawrence, Lee DuBridge, I. I. Rabi, Th. von Karman, W. R. Sears, L. Berkner, A, Compton, M. A. Tuve, and C. C. Lauritsen. These men served on the important scientific advisory committees of the Navy, Army, Army Air Force, and the Atomic Energy Commission (AEC) (4). They helped to determine policy and the areas of research that were to be supported. They were the links between the military and the leading academic institutions, for they were all respected members — in fact, the intellectual and administrative leaders — of the premier institutions in the country. They deeply influenced the character of the alliance and in important ways shaped the institutional framework and the directions in which science developed in the U.S. in the post — World War II period.

Many of these scientists were physicists. This reflected the special role that physicists had played in the development of radar and the atomic bomb: not only had they displayed a mastery of the fundamental science governing the phenomena, but they also had proved themselves to be talented gadgeteers and engineers, as well as able production managers. Physicists had been concerned with every facet of the development of radar, from designing magnetrons and klystrons to overseeing the manufacture of the completed sets, including antennas and scopes. Shortly after the war, Lee DuBridge described the transformation of the physicists who had came to the "Rad lab" at MIT, whose director he had been:

What happened to that group of academic physicists who had been so suddenly plucked from their ivory tower labs? Well, of course, by 1943 there were several hundred of them — and they were now being aided by hundreds more of engineers, mathematicians, astronomers, physiologists, lawyers, business men, to say nothing of machinists, secretaries, janitors, plumbers, carpenters and an uncountable number of guards. But it was always that initial group of 75 physicists who supplied the leadership. Maybe the prejudices of the Director had something to do with it. Maybe we were able to attract the best physicists and not the best engineers. But the physicists were just *good*. They were objective, they were imaginative. They jumped at new ideas, they were enthusiastic, genial, cooperative. They had a happy, exciting, wonderful time.

But they weren't really physicists any more. They weren't doing research for the fun of learning new things. They were developing weapons for war for use *tomorrow*. They had become engineers, military strategists, salesmen, production experts. And they were the most *travelling* group you ever saw. We quickly found that the only satisfactory way to exchange information rapidly and effectively was to do it person to person. Many of us seldom had a straight week at home. Some were gone for weeks or months at a stretch. They visited air bases, factories, the Pentagon, and the Navy buildings, other Army, Navy, and civilian laboratories in this country and in Canada, in England, in Australia. In 1943 we established a branch laboratory in England to keep in closer touch with British radar workers and to be of more direct help to our own fighting forces. A few days after the liberation of Paris we had a field station there . . . (5).

And the same was true in the development of the atomic bomb (6). Physicists had not only constructed the first experimental pile, designed the first uranium and plutonium bomb, but had also been deeply involved in the construction of the Hanford reactor and the Oak Ridge separation plant.

Physics seemed to hold the solution to the problems posed by the new weapons that had been developed during the war, particularly the problems connected with sensing devices, radar and communication, and nuclear weaponry. Physicists seemed uniquely well equipped and particularly eager to address these problems. Their eagerness stemmed from a host of reasons. They had tasted sin, and many of them believed that a strong United States implied that the weapons they had crafted would never be used again. The fight over the control of atomic energy had taught them that they must translate their moral convictions into concerted political action (7). In the new world they had helped create, they could no longer isolate themselves in their ivory towers from the affairs of the nation. Their moral duty as citizens implied service to the nation; and for many of the leading physicists this meant active participation in the huge enterprise that had been undertaken by the military to shore up the peacetime defenses of the country. The rebuilding and strengthening of the scientific establishment was part of that effort. It also had not escaped them that universities — and physics departments in particular — were major beneficiaries of that enterprise.

The period from the end of World War II to the beginning of the 1960s was a unique period in physics in the United States — the new Atlantis, where Bacon's College of the six Days' Work was being built. Essentially unlimited resources were being allocated to realize the end of that foundation: To apprehend "the knowledge of causes, and secret motions of things, and the enlarging of the bounds of human empire, to the effecting of all things possible" (8). The instruments created in this American House of Solomon were indeed Bacon's fantasy world reified: "high towers," "chambers," "perspective houses, where demonstrations of all lights and radiations [were made]," "means of seeing objects as far off, as in the heaven and remote places," "mathematical house[s]" with "exquisitely" made instruments. Bacon's Merchants of Light became scientific attachés in their modern incarnation. "Depredators," "Mystery-men," "Pioneers," "Compilers," and "Dowry-men" were to be found in both houses, attesting to some constant features of human nature. His interpreters of nature are our theoreticians, "those that raise the former discoveries by experiments into greater observations, axioms and aphorisms." As benefited their Baconian ancestry the scientists of this new house had "their consultations" to decide "which of the inventions and experiences which [they had] discovered shall be published and which not." But whereas in Bacon's utopia the Elders of Solomon's house had the right to conceal from the state the inventions they thought fit to keep secret, in the new Atlantis the scientists could not. Oppenheimer's downfall — the dominant political event in the life of the scientific community during the 1950s — made it clear that although scientists had powers, Power rested with the state.

The initial funding for this great postwar American enterprise came from the Navy, the Air Force, the Army, and the AEC (9). One of their aims was the development and maintenance of an adequate scientific manpower pool to meet the defense and technological needs of the United States. The support was so lavish that all needs were met: those of the universities, those of industry, and those of the state.

One of the distinctive features of the decade, the emergence of high-energy physics, was an outgrowth of this bounteous ministration by the state. It had been widely perceived that the success of the great wartime laboratories,such as the Rad Lab at MIT, the Met Lab at Chicago, and Los Alamos, reflected the ability of experimenters and theorists to collaborate intimately with one another and to work closely with industry — a tradition that had been fostered in the nuclear research laboratories established during the thirties at various American universities, and particularly at Berkeley in Lawrence's Rad Lab (10). One of the motivations for the funding of high-energy research was the maintenance of centers in which this kind of relation between theory and experiment, and of experimenters with industry, would continue to flourish in large-scale projects. Incidentally, Bell labs, for similar reasons, was the principal beneficiary of postwar military funding in the development of the transistor (11).

Government funds made available through the Office of Naval Research or the AEC paid most of the expenses for most of the high-energy machines that the universities acquired from 1945 on. By 1950 there were four large synchrocyclotrons in operation in the United States (at Berkeley, Rochester, Harvard and Columbia) and two under construction (at Chicago and Carnegie Tech). At the end of the decade, some twenty universities had major high- energy installations (12).

The "high-energy" theoreticians educated during this period were trained with close links to experimental practice; they were instilled with a pragmatic, utilitarian outlook and were taught to take an instrumentalist view of theories. This was the legacy of the approach that had proved so successful at Berkeley, Columbia, Harvard, and other leading American universities during the thirties. And the same was true of "solid state" theoreticians, for whom connections with applied physics had always been the norm.

Pragmatism was something constitutive of the American theoretical physics community. It was an imprint given it by the people who were responsible for its creation, its nurturing, and its growth in the period before World War II — namely, experimentalists such as P. Bridgman, K. T. Compton, C. B. Millikan, and H. M. Randall, theorists such as E. C. Kemble, G. Breit, J. H. Van Vleck, R. J. Oppenheimer, and J. C. Slater, and later by the like-minded "refugees" who were offered positions in the United States after Hitler's rise to power in Germany: H. A. Bethe, F. Bloch, E. Fermi, and E. Teller, among others. The pragmatic stance was reinforced by the institutional context in which it developed: departments of physics in which experimenters and theorists were housed together and plied their trade together under one roof, in contrast with the European situation (13). In nuclear physics, in particular, a symbiotic relationship grew between theorist and experimenter; partly because essentially no nuclear theory had existed before 1933, and partly because the explanation and understanding of the data coming out of the cyclotrons and β-ray spectographs required fairly recondite theory.

This "American" style of doing physics was charcteristic of the great wartime laboratories: the Radiation laboratory at MIT, the Met Lab in Chicago, and Los Alamos. It was here that many of the outstanding American theoreticians of the fifties were molded – for example, R. P. Feynman, G. F. Chew, M. L. Goldberger, R. E. Marshak, and J. S. Schwinger. It is a style that was institutionalized at all the leading departments during the fifties and became the national norm.

In the years immediately following World War II, from 1946 to 1950, funding from the Office of Naval Research and the AEC – the principal sources of support of physics in academic institutions – was determined by considerations that were primarily internal to the discipline, and there were no explicit demands attached to grants or contracts. Then the cold war, the detonation of the first Russian atomic bomb, the Korean war, and the McCarthy era polarized the political climate, and thereafter the character of the support changed. Although the research supported at universities by the Department of Defense (DoD) and the AEC remained for the most part unclassified and publishable, applied research began to receive a greater emphasis. National priorities, and defense needs in particular, began to determine more directly the areas and directions that were to be sponsored. The establishment of the National Science Foundation (NSF) in 1950 reinforced the tendency of DoD to support basic research only in those areas likely to benefit the military needs of the country. This change coincided with a new generation of officers taking over the affairs of the Pentagon (14). It also coincided with the strengthening of the older Army and Navy laboratories – such as the Army Ordnance and Signal

Corps laboratories and the Naval Research Laboratory — and the establishment of new advanced research facilities — such as Wright-Patterson, the Air Force Research Directorate at Hanscom Field, and the Lincoln Laboratories (15). Similarly, at Los Alamos and at Livermore the permanent staff took charge of nuclear weapons research and development. This was part of policy by the Armed Forces to regain control over all aspects of weapons research and development, control that had been taken from them by OSRD during the war. The hope of controlling military technology through civilian boards composed mostly of scientists, boards that would act autonomously and be free of political interference — Bush's dream — was shattered by the events leading to the decision to build the hydrogen bomb. The policy of direct scientific advice to the president — another Bush legacy — proved somewhat longer lasting. The dismantling of the President's Scientific Advisory Committee (PSAC) by Richard Nixon was a landmark; technology and applied science policy henceforth replaced science policy at the highest level (16).

But the value of the advice and talents of leading physicists was not forgotten. If, during the 1950s, PSAC recruited the outstaning established physicists, the AEC and the Department of Defense wooed the best and brightest of the post—World War II generation of physicists. The best young theoreticians were invited to work summers on various defense-related projects at Los Alamos, the Rad Lab in Berkeley, and other DoD- or AEC-supported laboratories; and later that decade, with the Institute for Defense Analysis. Defense projects are by their very nature applied, practical, and gadget-oriented. The tools in the theoretician's armory that are essential for this enterprise are often the same as those that render the theoretician valuable to his experimental colleagues in their laboratories.

The defense connection during the 1950s reinforced the pragmatic, utilitarian, instrumental style so characteristic of theoretical physics in the United States. The successes of this mode of doing theoretical physics help to explain its diffusion to Europe and elsewhere. The progamatic ideal of American physics which had been visible from early on, now became not only the national norm but, in fact, predominant worldwide.

The period under consideration witnessed the disappearance of the

Newton-like "geniuses," the "off-scale" creative indviduals who by *themselves* work out most of the details and consequences of their brilliant insights and ideas. Consequences and applications now were being worked out by a collective effort. Community and communal activity became key features in the activities of theorists, and particularly of high energy theorists. A multitude of factors — the greatly increased size of the theoretical community, the stiffer competition entailed by this, the pressure to account for the data being poured out by the complex and costly apparatus in the experimental laboratories, the pressure to suggest new experiments, the pressure to maintain funding — combined to result in the theoretical community's pouncing on new likely avenues and working out attractive suggestions as if it were a coordinated whole. It was as though most of the members of the community considered it more worthwhile to work out the approach suggested by the intellectual leader at the moment — for example, Gell-Mann, Mandelstam, or Chew in high energy — than to work on their own ideas or on longer-range programs of research. As early as 1951 Feynman called it the "pack" effect (17).

These developments were not, of course independent of one another. National manpower needs, defense needs, funding, the relation between institutional support and work on defense-related projects by the leading scientists of that institution, were clearly interlinked factors. The thesis is often asserted, especially by the practitioners, that context affects physicists but *not* physics. But recent historical studies relating to the development of the transistor, the laser, and quantum electronics more generally (18), suggest that in those fields where the foundations — quantum mechanics, and quantum electrodynamics — were secure, the thesis is not only open to question, it is false. To what extent and in what way this is also the case in the fields of physics where foundational issues are open — as with particle physics — is one of the challenging tasks confronting the historian of modern physics (19). The elucidation of the connection between funding, social and ideological constraints, and intellectual output is an important task. Insights will come from *detailed* studies of individuals and of institutions, their development and their interactions with one another and with the large context.

My intent in the present paper is principally to suggest questions. I will be concerned primarily with the Office of Naval Research, and its impact on the physics community in the period from the mid-forties to the mid-fifties.

Funding after the War and Forging of the Alliance

The United States emerged from World War II as *the* world power, and in possession of a weapon with which it believed it could impose peace on the world. The war had made clear that preparedness and scientific readiness, in particular, were going to be the most important factors in military power.

Vannevar Bush in 1944 noted that "if we had been on our toes in war technology ten years ago, we would probably not have had this damn war" (20). The allied victory was achieved by realizing the tremendous military potentialities that were latent in the scientific developments of the late 1930s. But as Fermi perceptively remarked in 1945,

> Nobody [in 1939] had any basis for predicting the size of the effort that would be needed and it may well be that civilization owes its survival to the fact that the development fo atomic bombs requires an industrial effort of which no belligerent except the United States would have been capable in time of war (21).

Similar, and perhaps more appropriate, remarks could be made regarding the development of radar, proximity fuses, bombsights, and the other devices that actually won the war, not to mention chemicals such as high octane gasoline or pharmaceuticals such as the sulfa drugs and atabrine. The United States' economic strength — the decisive factor in World War II — would no longer be as critical in the future. In any future conflict, long-range strategic bombers and rockets would deliver lethal blows at the outset, and the likelihood of a protracted war was slim. The United States would therefore not have time to convert its civilian industries to the mass production of weapons, as it did in World War II. Technological sophistication and readiness in *peacetime* would henceforth assume paramount importance.

In the past, the oceans separating the Americas from Europe and Asia had afforded the United States a great measure of security —

which meant that, for example, Pearl Harbor was attacked and not the continental United States. But rockets and long-range bombers dissipated this geographical advantage. In his Biennial Report of 1945, General George Marshall stressed that "it no longer appears practical to continue what we once conceived as hemispheric defense as a satisfactory basis for our security. We are now concerned with the peace of the entire world. And the peace can only be maintained by the strong" (22). The security of the United States henceforth would have to depend on a system of far-flung bases situated all over the globe, from which all movements of potential enemies could be monitored. Simultaneously, the realization that wars had become too costly in their consequences to be fought led to the adoption of the policy of deterrence — including the possibility of preemptive strikes — as the best way to avoid wars. This fear of sudden annihilation, the legacy of Pearl Harbor, was further magnified by the development of the atomic bomb. Testifying before a congressional committee in the winter of 1944, Bush asserted,

Today it is evident to all thinking people that the evolution of new weapons may determine not only the outcome of battles, but even the total strategy of war. . . . Tomorrow the impact of new weapons may be even more decisive. . . . It is imperative, therefore, that after this war we begin at once to prepare intelligently for the type of modern war which may confront us with the great suddenness sometime in the future (23).

The importance of scientific and technological preparedness and of military strength is a constant theme in all the addresses of the scientific and military leadership during 1945. "How shall we maintain order while [a] peace loving world is being built?" Arthur Compton asked in the spring of 1945. His answer was, ". . . by keeping ourselves strong and working for friendly relations with our neighbours. . . . If we are to retain our leadership it will be only through superiority in those things that make a modern nation great [and] foremost among those things is science" (24). Karl Compton in June 1945 asserted that through weapons research, "the United States should be able, with only reasonable attention and effort, to keep itself in a position too impregnable to give any encouragement to any would-be aggressor" (25). William Patterson, the undersecretary of war, informed Congress that peacetime research would provide "a strength that will foreclose the possibility of others attacking us" (26).

In fact, the much shorter time-scale of weapons development brought about by the involvement of scientists during the war, and the concomitant danger of obsolescence, made peacetime weapons research imperative. Already in late 1944 the secretaries of War and Navy communicated to the National Academy of Sciences their view that,

> to insure continued preparedness along far-sighted technical lines, the research scientists of the country must be called upon to continue in peacetime some substantial portion of those type of contributions to national security which they have made so effectively during the stress of the present war . . . (27).

They also recommended a continuing working partnership among the military, industry, and the universities (28) — a notion strongly advocated by the scientific leadership, in particularr by V. Bush.

In June 1945, the Research Board for National Security — a joint Army, Navy, civilian board whose function was to organize and administer a "forward looking, long-range program of postwar research in scientific matters pertinent to national security" — was established (29). The Research Board was the result of a recommendation by a special committee on postwar research that had been appointed by the secretary of war and the secretary of the Navy; it was chaired by C. E. Wilson, of General Electric , and was composed of high-ranking members of the Army and Navy together with F. B. Jewett (president of the National Academy of Sciences), J. C. Hunsaker (director of the National Advisory Committee for Aeronautics), Merrill A. Tuve (director of the Applied Physics Lab at Johns Hopkins), and Karl T. Compton (president of MIT). Essentially, the Research Board was to take over the functions of OSRD, which was to be disbanded as soon as its affairs could be terminated after the war. The new board was conceived as part of the framework that was to oversee the support of research by the government after the war had been won.

A comprehensive program for postwar scientific research and scientific education had in fact been submitted by Vannevar Bush to President Truman in the spring of 1945. This report, *Science: The Endless*

Frontier, had been written by Bush at the request of President Rossevelt, after lengthy deliberations by his associates. In July 1960, on the occasion of the tenth anniversary of its foundation, the National Science Foundation reprinted the Bush report. In his introduction to this new edition, Alan Waterman, the then director of NSF who had been a high-level administrator in OSRD during World War II, asserted that NSF was reissuing the book "not as an historical document, but as a classic expression of desirable relationships between government and science in the United States." But by then Bush's vision of the "desirable relationships" had been considerably modified, and NSF was responsible for but a small part of the support of science by the government.

The message of *Science: The Endless Frontier* was the importance of basic research. Bush asserted that

> basic research leads to new knowledge. It provides scientific capital. It creates the fund from which the practical applications of knowledge must be drawn. . . . Today it is truer than ever that basic research is the pacemaker of technological progress. . . . *A nation which depends upon others for its new basic scientific knowledge will be slow in its industrial progress and weak in its competition in world trade, regardless of its mechanical skill* (30).

Basic research would provide the foundation for continued economic growth and for the development of new products and new technologies. The report stressed that knowledge of the methods and techniques of *basic* research were the essential ingredients in the training of all research investigators, be they in industry, in government laboratories, or in universities.

Bush emphasized in *The Endless Frontier* that it was essential for the national security of the United States "that there must be more — and more adequate — military research in peacetime." He suggested that this could best be done throgh a "*civilian-controlled organization* with close liaison with the Army and Navy, but with funds direct from Congress, and *the clear power to initiate military research* which will supplement and strengthen that carried on directly under the control of the Army and Navy" (31). He urged that the government accept new responsibilities for promoting the flow of new scientific knowledge and the development of scientific talent, and he recommended the establishment of a National Research Foundation to implement his recommendations for the government's support of basic research. He concluded the summary of his report with the statement:

> On the wisdom with which we bring science to bear in the war against disease, in the creation of new industries, and in the strengthening of our Armed Forces depends in large measure on our future as a nation (32).

Bush's detailed plan for a National Research Foundation — which included a Division of National Defense — encountered opposition from many different quarters. Nonetheless, it became the blueprint upon which the postwar legislation dealing with scientific research and scientific education was based (33). Most of his recommendations for a National Research Foundation — except those for the administrative oversight, and for a division of military research — were eventually incorporated into the legislation establishing the National Science Foundation and the National Institutes of Health. However, it would take five years of legislative debate to arrive at a bill acceptable to the Congress, the executive branch, and the scientific community.

The Office of Naval Research (ONR)

When, in the spring of 1945, Roosevelt denied funds to the Research Board for National Security, a group of young naval officers working in the Department of the Navy realized that it would be a long time before a National Research Foundation could be enacted into law, staffed, supplied with funds, and opened for business. These officers had recognized that Allied superiority from 1943 on was based on the introduction of new technologies into warfare, and they had noted that these technological advances originated with scientists who had been doing basic or applied research in universities and industrial laboratories. Throughout the war they had pointed out to their superiors that the Navy depended and relied on scientific and technological progress to carry out its mission. They were able to convince the secretary of the Navy that an Office of Naval Research was needed in order to carry out for the Navy in peacetime the functions that OSRD had performed during the war (34).

Public Law 588 establishing this office was passed by the 79th Congress on August 8, 1946. Even prior to that date, the Office of Research and Inventions (ORI) of the Department of the Navy (the immediate predecessor of ONR) was giving

> active financial support to scientific research workers in all fields of science — including medicine — in all parts of the country. Its contracts emphasized "the fundamental nature of research", and [were] carefully designed to preserve the freedom of inquiry and action so essential to the spirit and methods of research work (35).

Captain R. Conrad, the director of the Office of Naval Research's Planning Division, in the spring of 1946 stated, "I believe that the services are acting with wisdom and judgement that this support will not degenerate into military control of scientists: and that a firm base is being laid for the National Science Foundation of the future" (36).

Why was it that the Navy took the lead in supporting basic science? Part of the answer lies in the fact that there existed a long history of Navy involvement with science. In a special message on the occasion of the twentieth anniversary of ONR, the chief of naval research enumerated some of the landmarks in that relationship: "the early establishment of the Naval Observatory, the Navy's assistance in the creation of the National Academy of Sciences, and the emergence from its ranks of A. A. Michelson, the first Noble [sic] Laureate in Science" (37).

During World War I the Navy was the first branch of the armed forces to enlist the services of scientists. In 1915, the secretary of the Navy created the Navy Consulting Board to provide "machinery and facilities for utilizing the natural inventive genius of Americans to meet the new conditions of warfare"; Franklin D. Roosevelt, the assistant secretary of the Navy from 1913 to 1920, was a member of that board, which included two representatives from each of twelve of the country's leading scientific societies.

The fact that during World II respected scientists — such as Lloyd Berkner, Gaylord Harnwell, Harvey Hall, Philip Morse — had been high-ranking commissioned officers in the Navy performing important tasks and had been attached to the staffs of admirals who were influential in determining Navy policy, also played an important role. Of particular importance in shaping the Navy's involvement in the support of basic science was J. C. Hunsaker, the chairman of the National Advisory Committee for Aeronautics and a member of OSRD. Hunsaker was a graduate of the Naval Academy. He was an executive of the Naval Construction Corporation until 1926; thereafter, he was associated with the Bell laboratories and MIT. While at MIT in 1926 he instituted the first university course in aerodynamics and aircraft design. In 1941, Hunsaker was influential in having Admiral Furer appointed the Navy's coordinator of research and development and in selecting the staff of the Office of the Coodinator. He was the mentor of the "bird dogs" who formulated the plans for the Office of Naval Research.

Undoubtedly, the Navy's rivalry with the Army, and its desire to have access to nuclear expertise and technology, were also factors in setting up ONR. But a study of the documents issued by the high-level committees advising the various branches of the armed forces on scientific and technological matters gives support to the Navy's claim that ONR's vision was broader, and that its initial support of science was less mission-oriented, than that of the other services (38).

The preamble to Public Law 588, 79th Congress, summarizes the function of the Office of Naval Research:

> to plan, foster, and encourage scientific research in recognition of its paramount importance as related to the maintenance of future naval power, and the preservation of national security; to provide within the Department of the Navy a single office, which, by contract and otherwise, shall be able to obtain, cordinate, and make available to all bureaus and activities of the Department of the Navy, world-wide scientific information and the necessary services for conducting specialized and imaginative research; to establish a Naval Research Advisory Committee . . . to consult with and advise the Chief of such office in matters pertaining to research (39).

By 1949, ONR could boast that

> the huge university research program of the Navy Department is the greatest peacetime cooperative undertaking in history between the academic world and the government. This significant educational and scientific experiment now embraces approximately 1200 projects in about 200 institutions with a total expenditure of approximately $20,000,000 a year. Nearly 3,000 scientists and 2,500 college and university graduate students are actively engaged in basic research projects in the many fields of vital interest to the Navy. These projects were neither requested nor assigned by the Navy. The original proposals were initiated by the investigators (40).

In establishing these peacetime ties with universities, ONR followed the practice that OSRD had followed for mobilizing science during World War II. Two types of contracts were issued: one an omnibus contract that supported large laboratories and all their activities, and the other the much smaller-sized contracts that supported individual scientists, their research activities, and their graduate students.

The funds expended by ONR for basic research were always but a small fraction of the Navy's total budget. In relative dollar expenditures, ONR itself "never counted heavily in the Navy budgets." In the mid-sixties, ONR's total yearly budget was less than five percent of the Navy's one-and-three-quarter-billion-dollar annual expenditure on research and development. Yet ONR's impact had indeed been large. Taking stock of the state of affairs in 1966 – at the vicennial celebration of ONR – Robert W. Morse, the then assistant secretary of the Navy for research and development, asserted that the agencies of the federal government and the universities had followed successfully the guidelines set out by Vannevar Bush's report of 1945:

We have developed tools of both an administrative and institutional variety. We have developed understanding and support for research, not only within Government but amongst the public.

Moreover, concrete measures of achievement are considerable. The universities have developed a new style of graduate education and advanced training in the sciences under the kind of support made available first by ONR and later by many other Federal agencies. It has given the United States universities world leadership in these respects (41).

The vast network of support that ONR established from 1946 to 1950 was characterized by informality, a stress on excellence, and the open exchange of information. Reviewing ONR's research policy in 1954, Emanuel R. Piore, one of the highest-level civilian administrators within ONR during its formative years, noted that

the philosphy has been to resist any establishment of formal machinery in order to coordinate, plan, etc. We act as an umbrella for inter-Bureau and inter-laboratory groups to meet. We move into technical areas that appear critical or warfare areas that require pulling together. It is the general feeling that this is the most effective way to operate, for no one acquires a vested interest or a career in coordination, and one always looks at the content of programs, rather than the mechanics of getting data on beautifully designed charts so that to the untutored eye it may appear that everything is rosy. The coordination, the pulling together, of the research and development program in this administrative procedure, imposes a thorough look into the content, which, in turn, requires a high degree of competence on the part of those carrying the burden. The ability of the Navy, through the Office of Naval Reserarch, to operate in this fashion stems in large measure from the scientific programs that are supported in laboratories and through contract, by providing information on what is best in science and technology, and providing contact with the best minds in these areas (42).

The decision to operate informally, with expert civilian scientists overseeing and evaluating the work being sponsored at universities, was a policy that had been recommended by the Naval Research Advisory Committee. Section 4 of Public Law 588 establishing ONR and the Naval Research Advisory Committee (NRAC) stipulated the function of the later. It authorized the secretary of the Navy to appoint up to fifteen people from "those persons in civilian life who are preeminent in the fields of science, research, and development . . . to consult with and advise the Chief of Naval Operations and the Chief of the Office of Naval Research." The secretary was also given authority to define the term of membership and specify meeting dates. The first meeting of NRAC was held on October, 14 1946, with W. J. Kenney, the assistant secretary of the Navy, as temporary chairman; also present were Fleet Admiral Chester W. Nimitz, Admiral D. C. Ramsey (vice-chief of naval operations), and Vice Admiral H. G. Bowen (chief of naval research). At a subsequent meeting the members agreed to meet three or four times a year to review the program of ONR, and in the interval between meetings to assist the Navy and the chief of naval research, either as individuals or as subcommittee. At its formal meetings the NRAC received reports on the Navy's research efforts and considered the policy questions uppermost in the minds of those administering the research programs. Gordon K. Bell, the executive secretary of the NRAC, outlined some of the problems discussed at these meetings:

the advisability and practicability of hiring German scientists for work in this country; the Navy and the National Science Foundation; the possible transfer of funds from AEC to the ONR nuclear physics program; the supervision and coordination of research and development by the Office of the Secretary of Defense; Marine Corps research; Project HARTWELL; the wartime organization of science and the maintenance of basic research; contract policy; research and development on critical materials; untilization of scientific resources on a long-range basis and other classified problems (43).

After each of its formal meetings the committee was in the habit of meeting with the chief of naval operations and representatives from their offices. These briefings provided an opportunity "for a review of the day's discussion and the presentation of recommendations." The list of the civilian members of the NRAC in the interval from 1946 to 1954 includes Lowell T. Coggeshall, Karl T. Compton, Hugh L. Dryden, Leonard Carmichael, Lee A. DuBridge, William V. Houston, William S. McCann, Clark B. Millikan, William A. Noyes, Jr., J. Robert Oppenheimer, Lewis L. Strauss, Detlev W. Bronk, Richard J. Dearborn, Warren Weaver, Phillip M. Morse, and Arthur H. Compton.

Scientific Manpower

Much of the initial funding of ONR in basic research went to replenish the stock of scientific manpower of the nation. The leaders who had organized the American scientific war effor — Bush, Conant, K. T. Compton, Jewett — had been haunted by the fact that "in fundamental science, in particular, and to a large extent in applied science also, [the United States had] lost irrevocably the better part of a generation of creative research men and the better part of a generation of creative additions to [its] stockpile of fundamental knowledge" (44). They noted that the cadre of highly trained researchers in fundamental and applied science had not increased during the war, and that its average age was four or five years greater. Moreover, they believed that its creative potential had been diminished. They were also aware that there was a complete absence of fully trained men in the next lower age group — although this bracket was filled by those who had been partially trained when they went into war work. The war had resulted in a dearth of men who would normally have completed their undergraduate training and been candidates for advanced training. The Moe Committee — one of the task forces organized by Bush in the winter of 1944 to answer the question, "Can an effective program be proposed for discovering and developing scientific talent in American youth so that the continuing future of scientific research in this country may be assured on a level comparable to what has been done during the war?" — issued its findings in early May 1945 in a report entitled "Gather the Spilled Seed Corn." Its opening statement was:

The greatest cost of the war, in term of high living standards, full employment, and improved health for the American people in the future, has been the interruption of scientific training of men from 17 to 25 years of age – the most critical years. Whatever plan may be established for the more settled future, it is most urgent to take immediate steps for the recovery of scientific talent which has been diverted in its formative stages. . . . While our objective is peacetime prosperity, it may be valuable to recall that nearly all the fields with which we are concerned have a special value to national security (45).

Moe and his committee took the opportunity in their report to disparage those draft boards that, as the military situation improved and the need for war research leading to future weapons became less apparent, felt that advanced students who had been deferred for doing vital war research "should take their turn in the trenches":

We would point out . . . that the nation's interests are best served by adherence not to the concept of "equality of sacrifice" but rather that of "equal obligation to render the particular needed service that each can render best."

Moe went on to recommend that, even as the war was still being waged, "men now deferred for war research be enabled, as the urgency for such work diminishes, to resume their specialized studies and take up teaching duties" (46).

By the winter of 1944 the need to revitalize and strengthen American physics had become a major concern within government circles and within the physics community. The American Institute of Physics, which earlier in the war had become "the focal point for urgent questions and requests from agencies responsible for the conduct of the war" (47), in the spring of 1945 applied for and received a $29,3000, three-year grant from the Rockefeller Foundation to have its War Policy Committee (henceforth to be known as the Policy Committee) address the postwar challenges, particularly the physics manpower problem.

At the hearings to establish the Office of Naval Research, Admiral Bowen, who was to become the first chief of naval research, stressed that one of the goals of ONR was the development of a pool of skilled scientific personnel that was to be both larger and more adequately trained than that which existed before World War II — in case the nation needed to mobilize for war again. Bowen indicated that "most of the contracts are with universities . . . and the work done in many cases may be accomplished by graduate students who become acquainted with naval problems and may form a future scientific group to assist the Navy" (48).

In the process of rebuilding the nation's educational institutions, ONR initiated and supported many new branches of basic physics research in the United States — among them, radio astronomy, microwave spectroscopy, high-energy physics, and cryogenics. It also helped strengthen the fields of nuclear physics and solid-state physics. Similarly, basic research in almost every field of chemistry was sponsored by ONR, "though emphasis [was] given to areas not under intensive development by agricultural, industrial, medical and other interests" (49). The size of the basic chemical program in the 1946 — 1954 period can be estimated from the fact that about five percent of the papers published in American chemical journals in 1952 were based on work sponsored by ONR. Grants were given to encourage research in teaching institutions where little research had been carried out previously. Thus, the University of Rhode Island's Graduate School of Oceanography expanded dramatically after World War II, primarily as a result of ONR funding (50).

Another concern was the building of a cadre of navy officers with a scientific training. In 1946 ONR sponsored the establishment of a new advanced scientific curriculum at the Naval Postgraduate School at Monterey, California. This became a continuing program "to prepare selected officer personnel to deal with the problem of fundamental and applied research in the fields of general physics, chemistry, metallurgy, and applied mathematics" (51). Qualified naval officers were sent to graduate schools throughout the country to obtain advanced degrees (some 1,200 did so in 1952!); others matriculated in courses given at Monterey. These officers formed the nucleus of the scientifically trained personnel in the Regular Navy who, in the early fifties, assumed responsibility for and began overseeing the Navy's scientific and technological requirements in weapons and weapons research.

The rebuilding of the United States' scientific manpower base and the revitalization of basic science were primary considerations in ONR's funding policy immediately after World War II. But by 1954 ONR's concerns were different, and E. Piore informed the scientific community that

the scientific research of the Department of the Navy should provide a flow of fundamental knowledge of the sort the Navy needs in connection with present and future weapon developments; maintain contact with the scientists in the country in order to encourage their interests in fields of potential importance to defense; make the military services aware of new scientific findings; and provide a base for prompt mobilization of the scientific effort in case of severe emergency (52).

Piore also confirmed that when NSF came into being "the needs of the Navy gradually changed,"and that these were further modified when the Army assumed certain responsibilities by creating the Office of Ordnance Research, and when the Air Force did likewise by creating the Office of Scientific Research. Moreover,"this was according to plan." Piore also outlined the ground rules that governed support by ONR of programs in scientific research in the mid-fifties. The programs were classified into three general areas:

(a) Scientific areas that are of vital importance to the Navy with regard to exploitation for future readiness of our naval forces and which have at this time limited interest for exploitation to other components of the nation's community. Some examples are oceanography, Arctic research, upper atmosphere research, numerical analysis, applied mathematics, high energy (explosive) chemistry, and fluid mechanics. In these areas it is necessary at times to stimulate interest among the scientific and technical community. The military departments in these cases are usually the only source of such stimulation.

(b) Scientific areas that are of vital importance to the progress of naval technology and operations, and at the same time of great significance to industry and other components of society. Example of such fields are solid-state physics, statistics, microbiology, physiological psychology, electrochemistry, meteorology, and personnel training and selection.

(c) The last scientific area deals with those fields that at present appear to have comparatively remote possibilities as a source for naval exploitation, but yet there is sufficient potential to warrant modest support in order to ensure awareness of these areas and rapid exploitation when the occasion arises. This listening post activity must be of the highest quality and in balance with the rest of the effort (53).

By 1966, the goals had been narrowed still further. The then assistant secretary of the Navy for research and development, in an address on "Basic Research and Long-Range National Goals," declared:

> Let us be clear that the support of science for its own sake [i.e., basic research] is not quite the same as the support of scientists for their own sake. It is wise for society as a whole to support pure scientific research, not so much because it is an intrinsic virtue but primarily because it is an operational virtue. ... Academic freedom and the independence of pure research are special protections from arbitrary interference which a wise society promotes and defends for its own long-range good, not for the privilege of the recipients. ... Scientists, as special defenders of the faith, should not imply, ... that science is threatened if we care about its use ... (54).

The MIT Story

When the war ended, Los Alamos, Argonne, and the other installations that were part of the Manhattan District Project continued to be funded by that branch of the Department of the Army. Although OSRD had made provisions to extend the contracts that had supported the major wartime laboratories under its supervision — for examples, the Radiation Laboratories at MIT and Columbia — it made it clear that this support would be continued for a limited time only. As early as 1944, universities — in close contact with government agencies, and with their active support — began to formulate plans to salvage the wartime laboratories after the cessation of hostilities and to reconstitute themselves for peacetime activities.

Let me here briefly indicate the story of MIT (55). (Similar stories can be told for Berkeley, Columbia, Chicago, and elsewhere. By the spring of 1945, plans had been formulated to establish an interdisciplinary Research Laboratory of Electronics — with J. A. Stratton as its head, and with support from the Army, Army Air Force, and Navy — to continue in the postwar period the research activities the Radiation Laboratory had carried out during the war. A Division of Basic Research was also set up to consider other, more general, activities. Intensive discussions took place within the Department of Physics to plan for expansion in order to accommodate the expected large influx of graduate and undergraduate students in physics. Particular consideration was given to the problem of attracting (and supporting) the somewhat "older" young men, many of them with families, who had worked at the wartime laboratories with only a BS degree and had gained considerable experience there; the position of Research Associate (paying 3/4 of their wartime salary) was redefined to meet this problem. By the fall of 1945, 25 research associates had been appointed and 90 graduate students were enrolled; by the spring term of 1946, there were over 125 physics graduate students working for advanced degrees in the Department of Physics.

In the fall term of 1945 the Laboratory of Nuclear Science and Engineering (LNS&E) was established, with Jerrold R. Zacharias as its head, to oversee research activities and opportunities in the nuclear sciences. The initial stimulus for the establishment of such a laboratory came from Captain Conrad of ORI, who had contacted Dean of Science George R. Harrison and John C. Slater, the chairman of the department of physics. Interest in the field of nuclear physics and engineering was of predictable importance to the Navy. The objectives of LNS&E were as follows:

(a) To construct and make available to the Departments of Chemistry, Physics, Metallurgy, Biology, Chemical Engineering, Electrical Engineering, and Mechanical Engineering the necessary facilities for modern research in nuclear physics;

(b) To undertake new research — carried out in the various departments — which would contribute to the development of nuclear science;

(c) To train competent personnel in nuclear science and engineering able to design, operate, and use nuclear devices.

A number of scientists accepted Zacharias's offer to come to MIT, attracted in part by the promise and scope of the new LNS&E and the prospect of ample ONR funding; among them were V. F. Weisskopf, B. Rossi, H. S. Bridge, M. Sands, R. W. Williams, R. W. Thompson, G. E. Valley, and I. Getting. The budget of the laboratory for 1946 was about $300,000, most of the funds coming from ONR. Thereafter the annual budget climbed to 1.2 million dollars and remained fairly steady from 1948 to 1958. In 1958, after the launching of Sputnik, the AEC became the prime financier for the laboratory and its budget was expended.

The valuable resources that LNS&E offered to the Department of the Navy are indicated by listing some of the defense projects undertaken by highranking members of the laboratory. In 1946, under the auspices of the laboratory, Clark Goodman, then an assistant professor of physics, gave a *classified* course in nuclear reactor design (56). In 1947, LNS&E offered the first university-based course in reactor physics; one of the by-products of this course was the first textbook on reactor design and operation (57). These activities led to the establishment of the Department of Nuclear Science at MIT a few years later. In April 1948, C. Goodman helped initiate a high-priority program, directed by Captain H. G. Rickover, for the design of nuclear-powered submarines and the training (at LNS&E) of personnel to man such vessels. The first nuclear reactor was designed and constructed for the production of plutonium. The first power reactors were designed and constructed to provide propulsion for the Navy submarines. Incidentally, the subsequent developments of nuclear power — including reactor design — were deeply affected by the work at LNS&E.

The history of the development of power reactors illustrates a process that was to take place in other areas also. The training of a cadre of nuclear engineers, and the establishment of a nuclear industry operating under AEC supervision and supported by Navy contracts, effectively gave the military initial control over a large part of the nuclear enterprise. It marked the first stage of the dissolution of the physicists' dream of scientists as the overseers of the whole of nuclear technology. Similar histories can be traced in the fields of electronics, radar, transistors, lasers, etc.

In 1950, with J. Zacharias as director, LNS&E organized a summer study, in which some thirty scientists and engineers participated, to find ways to meet the threat that Russian submarines posed to overseas maritime shipping. Important suggestions for combating this danger were made at that time, which resulted in the development of lofar (low frequency directional listening arrays), in the use of helicopters for antisubmarine warfare, and in the construction of high-speed (Mariner) freighters. Zacharias was one of the group leaders of project Charles which led to the establishment of the Lincoln Laboratories (58). He was also techical director of Project Lamplight, which was established at Lincoln Laboratories in 1954.

But probably the most important dividend yielded by LNS&E was the personnel trained there. Between 1948 and 1958, roughly 400 persons participated in research at LNS&E, as part of the teaching staff or as graduate students or assistants. Of the 300 or so whose careers could be traced, it was found that in 1980 some 117 were professors in 38 U.S. and 10 foreign universities, 106 had key positions in 74 industrial corporations, and 57 were in government service. From 1946 to 1958 some 78 Ph.D.'s were awarded, and more than 1.000 undergraduates did their research as their senior thesis in the laboratory (59).

The funding by ONR extended to other projects at MIT; during 1946—1947, some thirty were being supported. One of the largest was Project Whirlwind: this project's original mission was the development of a new high-speed training system, a universal flight trainer-aricraft simulator, which would allow a pilot to "obtain a feel for flying a particular airplane before it was ever constructed" — but the objectives changed during the project to the development of a general-purpose computer (60).

Centers similar to LNS&E but with somewhat different interests and orientation were established at other campuses. All these "nuclear" laboratories had one feature in common: all of them (and in particular, LNS&E at MIT, the Newman Laboratory for Nuclear Studies at Cornell, and the Nuclear Lab at Chicago) attempted to re-create the atmospherics of Los Alamos — the sense of community, the exhilaration and joy in doing Physics — and to structure themselves so that they would be doing physics the way it was done at Los Alamos, with experimenters and theorists working closely together and interacting closely with one another. One further consequence of Los Alamos was apparent at these laboratories: theoreticians and experimentalists had become peers. The ultimate democratization of physics had been effected. In the thirties this democratic ideal had been realized by having experimentalists and theoreticians all members of the same department (but with experimentalists almost always chairmen). At Los Alamos — where Oppenheimer, a theoretician, was the guiding spirit — the theoretical physicists became invaluable and their powers became apparent. Theoretical physicists emerged from the war with a new status. Robert Wilson *and* Bethe provided the leadership at Cornell, Zacharias *and* Weisskopf at MIT, Fermi *and* Teller at Chicago.

Even more lavish support by the Atomic Energy Commission was responsible for the growth of the Radiation laboratory at Berkeley (61), of Argonne, and of Oak Ridge, for the establishment of Brookhaven (62), and for the construction of all the high-energy machines.

The scale of the support, its sources, and the requirements for its maintenance altered the character of science in the United States. What was meant by doing physics, was changed irreversibly.

Project Hartwell

In this section I want to focus on one of the many summer studies that MIT ran for the Office of Naval Research during the summer of 1950: Project Hartwell (63). It is generally agreed that this project – the first undersea warfare study undertaken after World War II – established a coherent perspective on all facets, technological as well as operational, that were found to affect the Navy's ability to safeguard the use of the seas in the presence of enemy submarines. The findings and recommendations of this summer study "gave structure and vitality to the Navy's ASW (antisubmarine warfare) program," and provided the guidelines for many of the subsequent developments: for example, the use of helicopters in ASW; the building of radars specifically designed for snorkel search; the improvement of acoustic surveillance of submarines; the possible use of atomic weapons in ASW; and the development of improved communications for submarines. The project was responsible for the development of the modern high-speed merchant vessels of the Marchant class (64). It also established the "systems" approach to defense problems.

The history of the project offers valuable insights into the relationship between the military, the universities, and individual scientists. The summer study was undertaken following a recommendation in January 1950 from the Committee on Undersea Warfare of the National Research Council to Admiral Momsen, the assistant chief of naval operations, that the threat posed by the new Soviet submarines be studied. The charge was wide-ranging and the study was to include: (a) transport and cargo handling; (b) vehicle and weapons systems; and (c) submarine defense. The committee recommended enlisting the assistance of a group of "outstanding personnel providing a high degree of technical, scientific, and operational competence under the most skilled and vigorous leadership." Momsen, together with Admiral Solberg (chief of the Office of Naval Research), met at the end of February 1950 with M. J. Kelly director of research of the Bell Telephone laboratories), his assistant J. B. Fisk, and J. A. Stratton (the provost of MIT). Stratton made it clear that MIT would cooperate with the Navy if the Navy felt that the danger posed by the Soviet submarines presented a critical national problem. It was the opinion of Fisk, Kelly, and Stration that J. R. Zacharias, the director of the Laboratory of Nuclear Studies at MIT, would be an excellent person to direct a group of scientists to study the long-range aspects of antisubmarine warfare. Admiral F. P. Sherman, the chief of Naval Operations, agreed with the recommendations of Solberg and requested the assistance of MIT in this matter. The president of MIT, Dr. J. R. Killian, then wrote to Admiral Sherman indicating MIT's readiness to undertake the project and on March 23, 1950, Zacharias was appointed director of the propossed group. The program was named "Project Hartwell (65). The plan of operation called for the period of intensive study to extend from June 5 to August 31, 1950. The thirty-three scientists recruited for the project included L. W. Alvarez (Berkeley), H. Brooks (Harvard), R. H. Dicke (Princeton), C. Eckart (Scripps), F. L. Friedman (MIT), I. A. Getting (MIT), A. G. Hill (MIT), F. V. Hunt (Harvard), C. C. Lauritsen (Cal. Tech.), P. M. Morse (MIT), A. Nordsieck (Illinois), J. A. Pierce (Harvard), E. Purcell (Harvard), and J. B. Wiesner (MIT). Several scientists associated with government and industrial laboratories also joined the project: P. Adams, L. V. Berkner, H. T. Friis, W. H. Groverman, W. E. Kock, R. K. Potter, R. B. Roberts, and F. L. Weldon.

During the first two weeks of June 1950, the members of the project were briefed on all aspects of the problem by high-echelon Navy personnel. The briefing included a visit to the Fleet Sonar School and Advanced Undersea Weapons School at Key West, observation of a day-long demonstaration of ASW in the Atlantic, a two-day visit to the New London Submarine Base and its submarine school, and a visit to the U.S. Navy Underwater Sound Laboratory. At the conclusion of the briefing on June 21 the group assembled at the Lexington Field Station of MIT, the home site of Project Hartwell. During July and August a large number of additional consultants visited the project, "to advise, to furnish information, and to comment on tentative conclusions." The Hartwell practice was to "summon" experts in any field under scrutiny or to send a small group to visit activities for information. A library of classified documents, secured through the assistance of ONR, was available to the project. Since only Zacharias, Friedman, Alvarez, and Lauritsen had Q clearance – the clearance necessary for access to materials dealing with nuclear weapons – they were the only ones to deal with these issues. A draft of the summary of the project's conclusions and recommendations was prepared in the final days of August 1950. It was discussed in detail at a terminal meeting on the 1st and 2nd of September that was attended by the Hartwell Group, high-level members of the Navy (including D. A. Kimball, the undersecretary of the Navy; Admiral F. P. Sherman, chief of naval operations; and Vice Admiral F. S. Law, chief of naval logistics), high-level personnel from the Office of Naval Research, Killian and Stratton of MIT, Kelly and W. B. Shockley from the Bell Telephone Laboratories, as well as G. P. Harnwell of the University of Pennsylvania, M. G. White of Princeton University, and F. W. Loomis of the University of Illinois.

The conclusions and recommendations of the report were:

(1) That additional atomic weapons for tactical use could and should be developed. The report stressed that "it is more important to develop light atomic weapons of small diameter, which can be delivered by small, fast, carrier-based aircraft. . . . Such weapons should prove useful against a great variety of military targets – for example, against ships, against shore installation, and against inland targets. Effective anti-submarine warfare requires above all the knocking out of enemy port installations. For use against port installations, both air-bursts and underwater explosions should be considered."

(2) That underwater atomic weapons be used against submarines, because they provide a mean of a sure kill when identification is possible.

(3) That deep-water test explosions be undertaken to determine the efficacy of atomic weapons in underwater use.

(4) That the rate of production of fissionable materials be increased immediately as a larger stock of atomic explosives was highly desirable.

Many other recommendations were included in the report, such as the use of helicopters in ASW, the construction of fast merchant ships with sea speeds greater than 19 knots, the dispersal of U.S. port facilities, the introduction of modern mechanical materials-handing equipment in U.S. ports to reduce turnaround time, and the improvement of radar and sonar equipment.

The stress on tactical atomic weapons was one side of an ongoing debate concerning the development of the hydrogen bomb. It reflected the position taken by Oppenheimer in that debate. Incidentally, on the eve of assuming the directorship of Project Hartwell, Jerrold Zacharias, together with J. R. Wiesner and colleagues from MIT and Harvard, drafted a lengthy statement published in the *Christian Science Monitor*, the *New York Times*, and the *Washington Post* deploring the present military policy, which had produced "an overwhelming dependence on atomic weapons." The letter asserted that policy "was founded on misplaced faith in atomic weapons and strategic bombings" and that the United States must "move beyond central reliance on atomic warfare and must accept the costs in money and inconvenience involved in such a shift in policy" (66).

From the point of view of physics, the most interesting contribution of Project Hartwell is contained in Appendix J of the report, which was the work of R. H. Dicke and E. M. Purcell, both former members of the MIT Rad Lab. It outlined the construction of a magnetometer for magnetic airborne detection using electron spin paramagnetic resonance absorption. The discussion made elegant use of dispersion relations, in what must have been one of their first uses in solid state theory. Dispersion relations are by their very nature "phenomenological"; they relate the real and imaginary parts of various physical quantities by virtue of causality conditions (67). Appendix J illustrates the thesis that the involvement of many of the leading American physicists in defense matters reinforced a theoretical orientation — pragmatic, phenomenological — that had become the norm in the country.

There is little question that funding affected the development of solid state physics and what later became known as condensed matter physics. Did funding also affect the cognitive aspects of theoretical high energy physics, that is, foundational physics? Surely the fragmentation of the interests of the leading theorists (stemming from their consulting and their involvement in defense matters) hindered — and to certain extent prevented — their maintaining a sustained focus and effort on fundamental theory. Further, in their capacity as reviewers of research proposals, and by virtue of their dominance in the funding process, they tended to reinforce their dominant view, which was pragmatic and phenomenological. At the very least, I would argue that until the early sixties the *tempo* of developments and the rate of initiation of new viewpoints in the field was affected.

Linkages

Another aspect of the Hartwell project should be noted. The program came into being primarily at the instigation of civilian scientists (such as G. P. Harnwell) who had worked on submarine warfare during the war, and were presently members of the NRC Committee on Undersea Warfare. Harnwell, the deputy chairman of that committee, had been a commander in the Navy and had served on the staff of the Navy's submarine command. Once the problem had been perceived and brought to the Navy's attention, a network of contacts allowed the chief of naval operations to turn to MIT and set into motion the events that led to the summer study. Project Hartwell was made possible by the existing linkages. This summer study reinforced them and established new ones. This pattern would repeat itself often during the fifties (68).

The war had created a hierarchical, pyramidal structure of the scientific community insofar as relations with the centers of governmental power were concerned. At the apex of the hierarchy had been Bush and Conant, with K. Compton and Jewett close behind. But by the late forties a new generation had taken over the leadership: among them; Rabi, Stratton, Lawrence, DuBridge, and Oppenheimer. They were influential by virtue of the respect they commanded in the scientific community, and by virtue of their administrative positions. It was to these people that the Department of Defense turned for advice and for recommendation of scientists to staff committees and to direct special projects. Thus, Stratton recommended Zacharias, who in turn recruited the scientists for the project. For the final evaluation of Project Hartwell, Harnwell, Loomis, and M. G. White were brought in. Harnwell at the time was chairman of the large Department of Physics at the University fo Pennsylvania, and a few years later would become the president of that institution. M. G. White was in charge of the Princeton cyclotron, and a few years later would become director of the Penn-Princeton synchrocyclotron, a project largely funded by ONR. Loomis was chairman of the Department of Physics at the University of Illionis, one of the largest departments in the country, and an influential and respected member of the council of the American Physical Society, who would become the first director of the Lincoln Laboratories.

The involvement in the project of such first-rate scientists as Alvarez, Brooks, Morse, Purcell, and the others, not only guaranteed highly valued advice, suggestions, and recommendations, but also resulted in technical breakthroughs that then were exploited and worked out by the staffs of various government and industrial laboratories. Another facet of their involvement ought to be stressed. These outstanding scientists helped create a social reality in which working on military problems was an accepted norm for scientists at universities; more than that, their involvement suggested that one aspect of being the very best was participation in defense project. The physics departments at the premier schools (MIT, Chicago, Princeton, Harvard, Berkeley, etc.) thus became institutions in which many of the best graduate students conceptualized the political world in terms of the model presented by these outstanding scientific workers. That construction of social reality was shattered only during the Vietnam war.

With the chilling of the cold war — particularly after the explosion of the first Russian atomic bomb in 1949, and the outbreak of the Korean war — the AEC intensified its research activities in fusion processes. Many of the senior theoreticians who had worked on the atomic bomb project participated in these efforts. Many of the younger theoreticians were invited to join projects at Los Alamos, at the Radiation Laboratory at Berkeley, at Princeton, and elsewhere, to work either full time or during the summer, and many accepted. By the middle of the decade the value of such contributions was clear, and the secretary of defense and the Joint chiefs of Staff made a request for an organized university effort to support and strengthen the Weapons System Evaluation Group of the Department of the Defense "as a public service." The Institute for Defense Analysis (IDA) was created to fulfill this request. It is an association of universities that was formed "to promote in the field of Defense studies, a more effective relationship between the national security and scientific learning" (69). Incorporated in April 1956 and supported by a Ford Foundation grant of $500,000, IDA was to be a link between the scientific and technological community and the Department of Defense. Its first annual report, issued in 1957, gave the rationale for its mission:

> During World War II the process of converting the discoveries of science and the developments of technologies into new elements of military power — including the necessary readjustments in military strategy and tactics — went at an undreamed of pace. Developments in weapons and military equipment found their way so quickly into the pattern of military operations that the whole shape of military power was radically changed in the span of half a dozen years.
>
> Radar, which at the time of Pearl Harbor was a rudimentary and untrusted device for the detection of aircraft in flight, had become a primary component of all major weapon systems by 1945. New aircrafts with vastly improved navigation and bombing systems contributed to powerful new strategic offensive capabilities. Uranium which at the beginning of the war was a scientific curiosity, had by 1945 been made into a weapon to stagger the imagination.
>
> Then toward the close of World War II, further developments in the population of air vehicles were bringing the dawn of yet another era of warfare, of high speed jet aircraft and very high performance guided missiles.
>
> Present military capabilities based on these new technologies in hostile hands present our country with a threat that is entirely unfamiliar: heavy destruction by direct attack. Moreover, the era of war in peace in which vast shifts in the world powers, aggravated by ruthless Communist ambitions of world domination, have brought us military responsibilities far beyond the direct defense of our own territory. . . .
>
> Such are the reasons why it is of paramount importance not only to give all necessary supports to weapons research, but also the maintenance of the most effective possible bridge between military strategy and the total of technology — for converting technical advances into new elements of military power — for guiding technology in the creation of new foundations of strategy.

IDA's purpose was "to assist in maintaining this bridge."

In 1958 (following Sputnik) IDA signed a two-year contract with the National Security Agency to extend its activities. In the summer of 1959 C. H. Townes, then at Columbia, wrote to Garry Norton, the president of IDA, that he and Marvin Stern had met with a group of scientists at Los Alamos (K. A. Brueckner, M. L. Goldberger, K. M. Watson, N. M. Kroll, and S. Treiman),

> to explore possible arrangements whereby more first rate scientists and perhaps highly trained professional persons may contribute effectively and creatively to the National Security. . . . A particular need . . . is good contact and communication with the younger established scientists who are in their most productive period.

The names suggested were K. Brueckner, S. Chandrasekhar, V. L. Fitch, R. L. Garwin, M. Gell-Mann, M. Goldberger, R. Karplus, N. Kroll, H. Lewis, D. Lieberman, C. L. Longmire, F. Low, D. Parker, F. Reines, M. N. Rosenbluth, S. Treiman, and K. Watson. Thus were many of the young Jasonites born.

All federal agencies seeking close ties with top scientists followed this same pattern. Willard Libby, the Nobel laureate chemist who served as an AEC commissioner during the Eisenhower administration, confided:

> You see, here's what you do: it's a kind of modern piracy. When I was on the AEC I carefully "ground in" all the top talent I could find in the physical sciences and we had nearly 99 percent of them. You see, you know who they are through your connections and you can get them, and you just hold them. And you give them everything they need and they get loyal to you and pretty soon you've got them, you see. This is what can be done, and this is the great battle among agencies (70).

Epilogue

Even though physicists remained respected members of the high-level boards advising the armed forces and the president, by the mid-fifties their special status had been eroded. Their unique skills as the makers of novel energy sources and advanced instrumentation were still much sought after — as in the case of the development of masers and lasers — but they no longer controlled the theory and practice connected with the new weaponry. Electrical engineers took over the field of communications systems, including antenna design, and they were primarily responsible for the development of computers; often they were indistinguishable from solid state or "applied" physicists.

The story of the physicists' association with the military in the period 1945—55 is a tragic one. They had assumed that the guidelines under which they had served the nation during the war would continue to be valid in their peacetime association with the military. They had believed that the role the military would play in the post—World War II world would remain the same as it had been before and during the war. The more perceptive among them realized that the rules and roles had changed during the early fifties. Others realized only much later — to their dismay and sorrow — that they had been used.

Acknowledgement

I would like to thank Edward Purcell for a valuable conversation, Myles Gordon for a helpful discussion, and A. Hunter Dupree for sending me a copy of the insightful article he published in *Nature* in September, 1986. The constructive criticisms and comments following the presentation of my paper at the conference were responsible for its improved, present version.

Notes and References

1. The literature on World War II and the scientific community is by now vast. One of the best points of entry is Daniel Kevles, *The Physicists* (New York: Knopf, 1978, particularly valuable is the "Essay on Sources" in that volume. See also A. Hunter Dupree, "The Great Instauration of 1940: The Organization of Scientific Research for War," in G. Holton (ed.), *The Twentieth Century Sciences* (New York: Norton, 1972), pp. 443—467; Carroll Pursell, "Science Agencies in World War: The OSRD and its Challenges, in Nathan Reingold (ed.), *The Sciences in the American Context: New Perspectives* (Washington, D.C.: Smithsonian Institution Press, 1979), pp. 359—379; Alex Roland, "Technology and War: A Bibliographic Essay," in Merritt Roe Smith (ed.), *Military Enterprise and Technological Change* (Cambridge, Mass.: MIT Press, 1985), pp. 348—379.
2. For an overview of the wartime and postwar developments, see A. Hunter Dupree, "National Security and the Post-war Science Establishment in the United States," *Nature* **323** (September, 1986), pp.213—216.
3. The partnership aspect should be stressed. It seems to have operated at all levels. I. I. Rabi tells the following story of his wartime experiences at the MIT Rad Lab: "We had to set relations with the military. I was very forceful about that myself.

For example, one group came from the Navy. They wanted certain black boxes, which they described, to be developed with certain voltages, and so on. I asked, 'What are they for?' Their exact answer was 'We prefer to talk about this in our swivel chairs in Washington.' I didn't say anything. Neither did I do anything — except continue to develop three-centimetre radar. They came back some six months later — the same thing. I said, 'Now, look, let's stop kidding. Bring your man who understands aircraft, and we'll talk about your problem.' By that time, we had learned a lot about radar and tactical military application. Well, the Navy did. We found that their problem was to knock off Japanese aircraft spying on ships. It turned out that they needed a shipborne height-finding radar to supplement and guide the radar equipment already in their carrier-based planes. And we made an agreement with the Navy. We'll develop that if you and we can do the whole thing together — a partnership. We're in this war together. We can talk about the whole thing, whatever it is, and then our side will do its best to develop the appropriate radar. Which it did. It was a fantastically great radar — a very effective thing. As time went on, we set up an effective pattern of interdependence with the military. Fortunately, our money did not come from the military directly but from another government agency, the office of Scientific Research and Development, under Dr.Vannevar Bush. Actually, we would use this money to develop a particular radar. We would then try to interest either our military men or the British in it. If they were interested, then they financed the production. After we learned to get along with the military men, we grew to have a deep respect for them — respect for their devotion and hard work. We got along with them once they saw that we were not there to take anything away from them but actually to help them" (in J. Bernstein, "Profiles, Physicist II," *New Yorker*, October, 1975). John Rigden in his biography of I. I. Rabi gives illustrations of this partnership; see John S. Rigden, *Rabi: Scientist and Citizen* (New York: Basic Books, 1987).

4. Many of them served on several of these committees simultaneously. Thus for example, Hugh Dryden and Lee DuBridge were members of the Army Air Force Scientific Advisory Group, at the same time as they were members of the Naval Research Advisory Committee. For a list of the committees that Rabi served on concurrently see his testimony in the Personnel Security Board investigating J. R. Oppenheimer: *In the Matter of J. Robert Oppenheimer* (Cambridge, Mass.: MIT Press, 1971), pp. 451—473.
5. Lee A. DuBridge, "The Birth of Two Miracles," The California Institue Forum 6, 1949, pp. 1—15. Edward Purcell, who was one of the original members of the Rad Lab scientific staff and helped design the first operational radar set during the winter of 1940, made the following observation: The British had conceived the mode of operation upon which the Rad Lab was patterned. However, the system did not operate quite as successfully there. "The British treated industry as tradesmen. Socially, the engineer was not treated as a peer by the academic physicist" (interview with E. Purcell, January 12, 1987). See also John Burchard, *O.E.D.: MIT in World War II* (New York: John Wiley, 1948); Phinney Baxter, *Scientists Against Time* (Cambridge, Mass.: MIT Press, 1968); and E. Pollard, *Radiation* (Duke Station, Durham: The Woodburn Press: Science and People, 1982). See also John Rigden, *op. cit.*, 1987 (3).
6. R. G. Hewlett and O. E. Anderson, *The New World 1939/1946: A History of the*

United States Atomic Energy Commission, Vol. 1 (State Park, Penn.: Pennsylvania State University Press, 1962); David Hawkins, *Project Y: The Los Alamos Story. Part I, Toward Trinity* (Los Angeles: Tomash Press, 1983); S. Groueff, *Manhattan Project: The Untold Story of the Making of the Atomic Bomb* (Boston: Little, Brown and Co., 1967).

7. See, for example, the address delivered in june 1947 at the commencement of the California Institute of Technology by its new president: L. A. DuBridge, "The Responsibility of the Scientist," *The California Institute Forum* **1** (1947), 1—15; see also Alice K. Smith, *A Peril and Hope: The Scientists' Movement in America, 1945—1947* (Chicago: University of Chicageo Press, 1965).
8. F. Bacon, *Essays and New Atlantis* (New York: Walter J. Black, 1942), p. 288.
9. Even a cursory examination of the sources for the support of the research published in the *Physical Review* during these years gives an indication of the overwhelming extent of the military and AEC support. See also A. Hunter Dupree, "The Structure of the Government — University Partnerships after World War II," *Bulletin of the History of Medicine* **39** (1965), 245—251.
10. J. L. Heilbron, R. W. Seidel, and Bruce R. Wheaton, *Lawrence and His Laboratory: Nuclear Science at Berkeley, 1931—1961* (Berkeley: Office for History of Science and Technology, University of California, Berkeley, 1981).
11. Thomas J. Misa, "Military Needs, Commercial Realities, and the Transistor, 1948—1958," in Merritt Roe Smith (ed.), *Military Enterprise and Technological Change: Perspective on the American Experience* (Cambridge, Mass.: MIT Press, 1985), pp. 253—288.
12. For a history of high-energy accelerators, see M. S. Livingston, *Particle Accelerators: A Brief History* (Cambridge, Mass.: Harvard University Press, 1969). For a contemporary history of the postwar development at Barkeley, see G. F. Chew and B. J. Moyer, "High Energy Accelerators at the University of California Radiation Laboratory," *American Journal of Physics* **18** (1950), 125—131. See also the review articles on high Energy Accelerators in the first two volumes of the *Annual Review of Nuclear Science*; in particular, see M. Stanley Livingston, "Synchrocyclotron" and "Proton Synchrotons," *Annual Review of Nuclear Science* **1** (1952), 163—168, 169—174; and E. L. Chu and L. I. Schiff, "Recent Progress in Accelerators," *Annual Review of Nuclear Science* **2** (1953), 79—92 (this article also contains an extensive bibliography of the literature until 1952). For the situation later during the decade see J. P. Blewett, "Recent Developments in Proton Synchrotons," *Annual Review of Nuclear Science* **4** (1954), 1—12; G. A. Behman, "Particle Accelerators: I. Bibliography, II. List of Accelerators—Installations," U.S. Atomic Energy Commission Document UCRL-8050 (1958), p. 153; *Proceedings of the CERN Symposium on High-Energy Accelerators and Pion Physics*, vol. 1, ed. E. Regenstreif (Geneva: CERN, 1956), pp. 9—11; D. L. Judd, "Conceptual Advances in Accelerators," *Annual Review of Nuclear Science* **8** (1958), 181—216. Behman, in his review article on accelerators (written in 1957), tabulated the date on accelerator installations throughout the world, listing information on a total of 500 machines in use or under construction; the *Proceedings of the CERN Symposium* (held in Geneva in 1956) gives information on some two dozen laboratories at which high-energy accelerators (>50 mev) existed or were under construction.

13. S. S. Schweber, "Empiricism Regnant: The American Theoretical Physics Community, 1918—1948," *Historical Studies in the Physical Sciences* **17**:1 (1986), 55—98.
14. See, for example, P. M. Morse, *In at the Beginnings: A Physicist's Life* (Cambridge, Mass.: MIT Press, 1977), pp. 245—261, esp. p. 249; this book offers valuable insights into the workings of high-level advisory committees in the post—World War II period, and documents the fierce interservice rivalries. P. M. Morse served on many influential committees, including the Weapons Systems Evaluation Group of DoD and the board of trustees of the Institute for Defense Analysis (IDA).
15. For a history of the Signal Corps at Ft. Monmouth, see Misa, *op. cit.*, 1985 (11). The history of NRL can be traced in J. A. S. Pitts, *The Historical Records of the Naval Research Laboratory* (NRL-MR-5109); see also David K. Allison *New Eye for the Navy: The Origin of Radar at the Naval Research Laboratory* (Washington , D.C.: Naval Research Laboratory, 1981); Louis A. Gebhard, *Evolution of Naval Radioelectronics and Contributions of the NRL* (Washington, D.C.: Naval Research Laboratory, 1979); B. C. Gibbs, *A History of the Chemistry Division NRL, Washington, D.C., 1927—1982* (Washinton, D.C.: NRL, 1983) (NRL-MR-5064). See also A. B. Christman, *History of the Naval Weapons Center, China Lake, California*, vol. 1, *Sailors, Scientists and Rockets* (China Lake, Calif.: Naval Weapons Center, 1971); N. A. Komons, *Science and the Air Force: A History of the Air Force Office of Scientific Research* (Arlington, Va.: Office of Aerospace Research, 1966); Arthur D. Tubbs, "Establishing Air Research and Development Command," Air Command and Staff College, 86—2570 (April, 1986).
16. For the hydrogen bomb controversy see, for example, Herbert York, *The Advisors* (San Francisco: Freeman, 1976); for PSAC, see William T. Golden (ed.), *Science Advice to the President* (New York: Pergamon, 1980).
17. R. P. Feynman, "The Present Situation in Fundamental Theoretical Physics," *Annales da Academia Brasileira de Ciencias* **26** (1954), 51—59.
18. The situtation for quantum electronics has been analyzed convincingly by Paul Forman, "Behind Quantum Electronics: National Security as Basis for Physical Research in the United States, 1940—1960," *Historical Studies in the Physical & Biological Sciences* **18** (1987), 149—229. I thank him for showing me a copy of this paper before publication. See also Paul Forman, "Atomichron: The Atomic Clock from Concept to Commercial Product," invited paper, *Proceedings of the IEEE* **73** (1985), 1181—1204. For the transistor, see Misa, *op. cit.*, 1985 (11). For the laser, see Joan Lisa Bromberg, "Engineering Knowledge in Laser Theory: Aspects of the Science-Technology Relationship" (to be published); Joan L. Bromberg, "Research Efforts That Led to Laser Development," *Laser Focus*, (October 1984), pp. 58—60. See also the article by David F. Noble, "Command Performance: A Perspective on Military Enterprise and Technological Change," in Merritt Roe Smith (ed.), *Military Enterprise and Technological Change* (Cambridge, Mass.: MIT Press, 1985), and David F. Noble, *Forces of Production* (New York: Alfred A. Knopf, 1984).
19. For the development of quantum mechanics, the relevance of context has been forecefully argued by Paul Forman, "Weimar Culture, Causality and Quantum Theory, 1918—1927: Adaptation by German Physicists and Mathematicians to a Hostile Intellectual Environment," *Historical Studies in the Physical Sciences* **3**

(1971), 1—115; see also Paul Forman, "Kausalität, Anschaulichkeit, and Individualität, or How Cultural Values Prescribed the Character and the Lessons Ascribed to Quantum Mechanics," in Nico Stehr and Volker Meja (eds.), *Society and Knowledge* (New Brunswick: Transaction Books, 1984). Marcello Cini has written on the relation between context and the development of quantum theory during the 1930s and 1950s: M. Cini, "The History and Ideology of Dispersion Relations. The Pattern of Internal and External Factors in a Paradigmatic Shift," *Fundamentae Scientiae* **1** (1980), 157—172; *idem*, "Cultural Traditions and Environmental Factors in the Development of Quantum Electrodynamics (1925—1933)," *Fundmentae Scientiae* **3** (1982), 229—253.

20. M. S. Sherry, *Preparing for the Next War: American Plans for Post-War Defense, 1941—45* (New Haven: Yale University Press, 1977), p. 128.
21. E. Fermi, "The Development of the First Chain Reacting Pile," *Proceedings of the American Philosophical Society* **90** (1946), 20—24.
22. G. Marshall, *Biennial Reports 1943—1945, War Reports*, quoted in Sherry, *op. cit.*, 1977 (20), p. 202.
23. Quoted in Moe Committee report, "Gather the Spilled Corn," American Institute of Physics. Barton Collection, Box 89, Folder 1.
24. A. Compton, "Science and our Nation's Furture," **101** (1945), 207—209. See also Frank B. Jewett, "The Future of Scientific Research in the Post-War World," in *Science in Progress*, Fifth Series (New Haven: Yale University Press, 1947), pp. 3—23.
25. K. C. Compton, *National Security and Scientific Research* (Cambridge, Mass.: MIT Press) 1945; text of a lecture delivered on April 19, 1945.
26. Sherry, *op. cit.*, 1977 (20), p. 129.
27. V. Bush, *Science: The Endless Frontier.* A report to the President on a program for post-war scientific research, July, 1945 (reprinted Washington, D.C.: National Science Foundation, 1960), p. 4.
28. Sherry, *op. cit.*, 1977 (20), p. 133.
29. *Science* **101** (1945), 226—227. See Also Kevles, *op. cit.*, 1978 (1).
30. V. Bush, *op. cit.*, 1960 (27), p. 19.
31. *Ibid.*, p. 6; emphasis added.
32. *Ibid.*, p. 9.
33. See, for example, *The Politics of American Science: 1939 to the Present*, revised edition, edited by J. L. Penick, Jr., Carroll W. Pursell, Jr., Morgan Sherwood, and Donald C. Swain (Cambridge, Mass.: MIT Press, 1972). The contract and grant system that had been devised by Bush at OSRD — the means by which the resources of universities were brought to the service of the nation without impinging upon the autonomy of the institutions themsleves — was one of his most important contributions to the postwar management of university-government relationships.
34. Bruce S. Old, "The Evolution of the Office of Naval Research," *Physics Today* **14** (1961), pp. 30—35. See also Harold G. Bowen, *Ship, Machinery and Mossbacks* (Princeton: Princeton University Press, 1954), pp. 30—35; Mina Rees, "Early Years of the Mathematics Program at ONR," *Naval Research Reviews* **30** (1977), 22—29; Harry M. Sapolsky, "Academic Science and the Military: The Years Since the Second World War," in Nathan Reingold (ed.), *The Sciences in the American*

Context: New Perspectives (Washington, D.C.: Smithsonian Institution Press, 1979), pp. 379—399; F. Weyl, "Research in the Service of National Purpose," *Proceedings of the ONR Vicennial Convocation* (Washington, D.C.: ONR, 1966), pp. 00—00; Edward I. Salkowitz, *Science, Technology and the Modern Navy: Thirtieth Anniversary, 1946—1976* (Arlington, Va.: Office of Naval Research, 1976).

35. Quoted in S. A. Glantz, "How the Department of Defense Shaped Academic Research and Graduate Education," in Martin L. Perl (ed.), *Physics Careers, Employment and Education* (New York: American Institute of Physics, 1978), pp. 109—122.
36. *Ibid.*
37. R. Adm. J. K. Leydon, "A Message from the Chief of Naval Research," *Naval Research Reviews* **XIX** (January, 1966), 1. Two documents indicative of the early involvement of the Navy in scientific matters are the "Report on the Erection of a Building in Washington as a Depot for Charts and Instruments" by Melville J. Gilliss and J. E. Nourse's "Memoir of the Founding and Progress of the U.S. Naval Observatory" (Washington, D.C.: 1871); both are reprinted in I. B. Cohen (ed.), *Aspects of Astronomy in America in the Nineteenth Century* (New York: Arno Press, 1980).
38. See, for example, *Science: The Key to Air Supremacy*, issued in 1946 by the AAF Scientific Advisory Group, Th. von Karman, Director. *Towards New Horizons*, the companion volume, includes the recommendations of the various subcommittees of the AAF Scientific Group: these dealt with Aviation Medicine and Psychology; Aircraft Power Plants; Radar and Communication; Explosives; Guided Missiles; Aircraft Materials and Structures. See also T. A. Sturm, "The USAF Scientific Advisory Board: Its First Twenty Years, 1944—1964," V/A Report, 1967.

    The Air Force Office of Scientific Research was officially established on August 11, 1955. It had existed, in recognizably similar form, since October 29, 1951. Its roots go back to the organization in February 1948 of a basic research group at Wright-Patterson Air Force Base.

    The Air Force began supporting "basic research" due to the stimulus of Theodore von Karman, the Hungarian-born aerodynamicist, who headed the Air Force's Scientific Advisory Board from its inception in late 1944. At the urging of the entire Scientific Advisory Board in 1948, the Office of Air Research was established within the Air Materiel Command, the organization charged at that time with Air Force R&D. The Office of Air Research was given a dual mission: to perform research in its own laboratories, and to conduct an extramural program by contract with industrial and university laboratories.

    In April 1951, following the establishment of the Air Research and Development Command — the first Air Force command organization wholly devoted to research and development — the Office of Air Research was split in two. The branch that remained at Wright Field was designated the Flight Research Laboratory, and was given an applied-research mission. The other branch was attached to Headquarters ARDC, in Baltimore, Maryland, and was given the job of running the contract program; it eventually became, in October 1951, the Office of Scientific Research (OSR).

    OSR was mainly concerned with administering a research program that, for the most part, was basic rather than applied, and was conducted entirely be means of contracts, mostly with university scientists. The research program was planned along the lines of the traditional scientific disciplines, and OSR itself was organized along these lines. There were originally five functional divisions — Chemistry, Fluid Mechanics, Mathematics, Physics, and the Solid State Sciences — each of which was headed by a division chief, usually a civilian, who was given considerable freedom in formulating a program for his division.

In February 1954, OSR's role in Headquarters ARDC was altered. OSR was stripped of its staff supervisory functions and was lowered a step on the organizational ladder. The overall supervision of ARDC research was put in the hands of a newly established Directorate of Research (Headquarters ARDC). In August 1955 the Office of Scientific Research became the Air Force Office of Scientific Research (AFOSR) and was given "center status," meaning that it was henceforth to be regarded, not as a headquarters staff section, but as an agency on the same footing with such ARDC field organizations as the Air Force Cambridge Research Center and the Wright Air Development Center. Reflecting the organization's new status, an Air Force officer, Brig. General Don B. Flickinger, was named its commander. See *Research Review, OAR*, IV, no. 6 (1965), 1—3.

39. The responsibilities of the Chief of Naval Research were spelled out in the directive issued by the secretary of the Navy to implement Public Law 588:

(a) The Chief of Naval Research shall be the principal adviser to the Secretary of the Navy on all research matters and such developments as may be expected from research.

(b) The Chief of Naval Research shall keep the Chief of Naval Operations advised of findings, trends, and potentialities in research and shall disseminate information to interested Bureaus and offices within the Navy Department, and to other Governmental or private agencies as may be appropriate, on naval and other research.

(c) The Chief of Naval Research shall be the principal representative of the Navy Department in dealings of Navy-wide interest on research matters with other Government agencies, corporations, educational and scientific institutions, and other organizations and individuals concerned with scientific research.

(d) The Chief of Naval Research shall survey the trends, potentialities, and achievements of scientific research and development and shall plan and coordinate research programs throughout the Naval Establishment.

(e) The Chief of Naval Research shall study and collaborate with the Chief of Naval Operations and the Bureaus in the formulation of the principal development programs of the Navy.

(f) All plans within the Navy Department for establishment of, abolishment of, or significant changes in, all laboratories concerned with research or development shall be referred to the Chief of Naval Research for comment and recommendation prior to final action.

(g) The Chief of Naval Research shall supervise, administer, and control all activities within or on behalf of the Navy Department relating to patents, inventions, trademarks, copyrights, royalty payments, and similar matters, and correlate such activities with the research and development activities of the Navy.

(h) The Chief of Naval Research shall undertake, (i) upon his own initiative, or at the direction of the Chief of Naval Operations, or at the request of any Bureau or office of the Navy Department, the development, design, maintenance, modification, and improvement of training devices and aids, and components thereof, and (ii) at the request of the Chief of Naval Operations or any Bureau or office of the Navy Department, the production of training devices and aids, and components thereof, to the extent that appropriations are made available therefor.

To discharge these responsibilities the Chief of Naval Research has available:

(a) The Naval Research Laboratory, for the conduct of basic research, applied research, and development.
(b) The Special Devices Center, for the development of training devices and aids, and related matters.
(c) The Patents Division, for the patent and royalties administration for the Department of the Navy.
(d) The Research Group, for the conduct of research by contract at universities, non-profit and commercial institutions, and other government laboratories. This Group also performs the staff functions of coordination, dissemination, planning, and evaluation of research, and related developments as assigned.
(e) The Branch Offices and Resident Representatives in the United States, to aid in administration of contracts at universities.
(f) The London Branch Office, to discharge the foreign responsibilities referred to in the Preamble to Public Law 588.
(g) The Underwater Sound Reference Laboratory, to act as a source of standards for underwater sound equipment and perform research in standards.

See E. Piore, "ONR Research Policy," *Naval Research Reviews* **VII** (April 1954), 6—11.

40. U.S. Navy, *Annual Report of the Office of Naval Research: 1949*, p. 1.
41. R. W. Morse, "Basic Research and Long-Range National Goals," *Naval Research Reviews* **XIX** (April 1966), 1—7.
42. Piore, *op. cit.*, 1954 (39).
43. G. K. Bell, "The Naval Research Advisory Committee," *Naval Research Reviews* **VII** (June 1954), 21—23. In 1954, the NRAC consisted of: A. V. Astin, Director, National Bureau of Standards; Luis de Florez, President, de Florez Engineering Company; R. E. Dyer, Director of Research, Emory University Hospital; J. C. Hunsaker, Department of Aeronautical Engineering, M.I.T., and chairman of the National Advisory Committee on Aeronautics; RADM P. F. Lee, USN (Ret.), Gibbs and Cox Co.; I. I. Rabi, Department of Physics, Columbia University; W. R. Sears, Director, Graduate School of Aeronautical Engineering, Cornell University; E. H. Smith, Director, Woods Hole Oceanographic Institute; H. N. Stephens, Vice-President, Central Research Department, Minnesota Mining and Manufacturing Company; J. E. W. Sterling, President, Stanford University; G. D. Stoddard, president of the University of Illinois. The committee was chaired by A. B. Kinzel, President, Union Carbide and Carbon Research Laboratories, and its vice-chairman was J. A. Hutcheson, Vice-President, Westinghouse Electric Corporation Research Laboratories.

44. Jewett, *op. cit.*, 1947 (24). See also Compton, *op. cit.*, 1945 (24).
45. Bush-Moe Committee, "Gather the Spilled Corn" (23).
46. *Ibid.*
47. H. A. Barton to Warren Weaver, April 6, 1949; attached to the letter is the report to the Rockefeller Foundation on the "Work of the Policy Committee 1945—1948." American Institute of Physics, Barton Collection, Box 89, Folder 1.
48. Quoted in Glantz, *op. cit.*, 1978 (35), p. 112.
49. L. W. Butz and R. Roberts, "Basic Chemical Research in ONR," *Naval Research Reviews* **VII** (July 1954), 1—4.
50. See, for example, "The University of Rhode Island's Graduate School of Oceanography," *Naval Research Reviews* **19** (September 1966), pp. 12—18.
51. B. I. Edelson, "Science Education for Naval Officers," *Naval Research Reviews* **VII** (September 1954), 20—22.
52. Piore, *op. cit.*, 1954 (39), p. 11.
53. *Ibid.*, pp. 11—12.
54. R. W. Morse, *op. cit.*, 1966 (41), p. 122.
55. See P. M. Morse, *op. cit.*, 1977 (14); James R. Killian, *The Education of a College President: A Memoir* (Cambridge, Mass.: MIT Press, 1985). See also Box 3 of the Morse Papers at the MIT Archives, MC75, which contains a history of the physics department at MIT.
56. *Nuclear Engineering at MIT: The First Twenty-Five Years* is a history of these activities; the volume was privately printed in 1982 by the Massachusetts Institute of Technology.
57. M. Deutsch, R. D. Evans, B. T. Feld, F. Friedman, and C. Goodman, *The Science and Engineering of Nuclear Power* (Cambridge, Mass.: MIT Press, 1948).
58. For a history of the establishment of the Lincoln Laboratories, see George E. Valley, "How the Sage Development Began," *Annals of the History of Computing* **7**, no. 3 (1985), pp. 196—226.
59. "Return on Investment in Basic Research — Exploring a Methodology," Report to ONR, Department of the Navy by Bruce S. Old Associates, Inc., November 1981, Contract N00014-79-C-0192. See also Rees, *op. cit.*, 1977 (34).
60. K. C. Redmond and T. M. Smith, *Project Whirlwind — The History of a Pioneer Computer* (Maynard, Mass.: Digital Press, 1980).
61. Heilbron, Seidel, and Wheaton, *op. cit.*, 1981 (10).
62. Allan A. Needell, "Nuclear Reactors and the Founding of Brookhaven National Laboratory," *Historical Studies in the Physical Sciences* **14** (1983), 93—122.
63. J. R. Marvin and F. J. Weyl, "The Summer Study," *Naval Research Reviews* **XIX** (August 1966), pp. 1—7.
64. Professor Zacharias made available to me a copy of the declassified report on *Project Hartwell, MIT. A Report on Security of Overseas Transport*, 2 vols. Contract No. N5 Ori 07846 (declassified by ONR Code 400, July 11, 1977).
65. After the Hartwell Farm Restaurant, which was located in Lincoln, at the edge of the MIT field station at Hanscom field where the summer study was to take place.
66. "Plea for a Closer Look at U.S. Security," *The Christian Science Monitor*, May 5, 1950.
67. See, for example, Charles P. Slichter, *Principles of Magnetic Resonance* (New York: Harper and Row, 1963), pp. 33—40.

68. A similar story can be told for the Air Force. On February 6, 1951, MIT initiated Project Charles, headed by Valley, a member of MIT's LNS. Project Charles (which became Project Lincoln) developed the plans for an experimental air defense system, and was responsible for the establisment of the Lincoln Laboratories of MIT; see Valley, *op. cit.*, 1985 (58).
69. From the Second Annual Report. All the materials relating to IDA have been obtained in Box 8 of the Morse papers at the Archives Institute, MIT (Morse was a trustee of IDA).
70. Quoted in D. S. Greenberg, *The Politics of Pure Science* (New York: The New American Library, 1967), p. 137.

PHYSICS BETWEEN WAR AND PEACE

PETER GALISON

Stanford University

Introduction

Three hundred and fifty years ago, Galileo introduced the notion of mechanical relativity by invoking the experience of sea travel:

> Shut yourself up with some friend in the main cabin below decks on some large ship, and have with you there some flies, butterflies, and other small flying animals. Have a large bowl of water with some fish in it. . . . The fish swim indifferently in all directions; the drops fall into the vessel beneath; and, in throwing something to your friend, you need throw it no more strongly in one direction than another. . . . You will discover not the least change in all the effects named, nor could you tell from any of them whether the ship was moving or standing still (1).

The imagery and dynamics of ships permeate Galileo's works, at once tying the new physics rhetorically to the modern navigational achievements of early seventeenth century Italy, and providing an effective thought-experiment laboratory for the new "World System".

Several centuries later, when Albert Einstein was struggling to overthrow Galilean-Newtonian physics, he too chose an image of contemporary transport as the vehicle for his radically new *Gedanken-experimente*. Now the railroads, symbol of the success of German technology and industry, replaced the sailing vessel in the argument. As Einstein put it, no optical experiment conducted in a constantly moving train could be distinguished from one performed at rest (2).

Yet a third technological image of transport appealed to the American physicist Richard Feynman as he assembled a synthetic picture of quantum mechanics and relativity in the years after World War II. The positron, as he saw it, could be viewed as an electron moving backwards in time. If so, then the simultaneous creation of an electron-

E. Mendelsohn, M. R. Smith and P. Weingart (eds.), Science, Technology and the Military, Volume XII, 1988, 47–86.

positron pair (ordinarily seen as involving two quite distinct paths) could instead be viewed as a single, continuous track: the positron travels backwards in time until it reaches the moment of creation, whereupon it becomes an electron moving forward in time. Feynman conveyed his vision with a vivid metaphor: "It is as though a bombardier flying low over a road suddenly sees three roads and it is only when two of them come together and disappear again that he realizes that he has simply passed over a long switchback in a single road." (3)

Every age has its cultural symbols, and Feynman's was as telling as Galileo's. Young American physicists of the 1940s and 1950s had seen their discipline recrystallize around the twin poles of radar and the atomic bomb. The venerated B-29 bomber carried both — an apt symbol, therefore, of the fruits of their labor — and served as a perfect vantage point from which to view the new theoretical and experimental physics.

Of course, the effect of the war on the development of physics in general, and of high-energy physics in particular, goes far beyond a passing metaphor. Indeed, the problem is that the effects are too great, and too varied, to be treated comprehensively in any one essay. For in a sense, the effects of the war permeate every aspect of postwar history. In the history of physics these effects include the thoroughgoing overhaul of the institutional structure of government-supported science. From the National Science Foundation to the Atomic Energy Commission, and the Office of Naval Research, no aspect of science funding remained unchanged (4). Among the war's consequences was a profound realignment of all relations between the academic, governmental, and corporate worlds, especially as physicists began contemplating the funding necessary for the construction of atomic piles, larger accelerators, and new particle detectors. Further, the war forged many collaborations and working groups among scientists that continued smoothly into the postwar epoch. And finally, the war provided astonishing quantities of surplus equipment that fed the rapidly expanding needs of postwar "nucleonics" — the study of a broadly construed nuclear physics, situated at the nodal point of research problems of cosmic rays, nuclear medicine, quantum electrodynamics, nuclear chemistry, and the practical imperatives of industry and defense.

Above all, one cannot ignore the new relation of university physics to military affairs that in a sense began, rather than ended, in the skies over Nagasaki. Suddenly academic physicists could negotiate with high-ranking officials from the Navy, the Air Force, and the Army to acquire new machines. At the same time, the military became an active participant in the shaping of postwar scientific research, through university contracts, the continuation of laboratories expanded during the war, and the establishment of new basic research programs under the aegis of individual armed services. Projects of joint civilian and military interest were lavishly funded, offering physicists the chance to think about exploring cosmic rays, not 3 or 4, but 100 miles above the earth's surface. Where a handful of technicians had once been sufficient to aid the physicists as they constructed new instruments, now the physics community began a deep new alliance with the various branches of scientifically informed engineers.

The present essay, addressing the impact of wartime research on postwar experimental and theoretical physics, can only begin to sketch some of these effects, drawing a few of the lines along which such a history of physics between war and peace might advance. I will not treat, for example, the vicissitudes in the dramatic careers of scientist-politicians such as Vannevar Bush, James Conant, or Robert Oppenheimer; the establishment and internal politics of funding organizations; or the alterations in industrial physics research policy. Instead, my goal is to peer into the effects of wartime science on the quotidian proceedings of physics itself, and into the experience of physicists in their research capacity. To do this I have chosen to focus on four exemplary physics departments: those at Harvard, Princeton, Berkeley, and Stanford. Each had its own trajectory, shaped in part by different war experiences and earlier patterns of research. Yet the four had much in common: each had to confront a sudden expansion, search for a new relationship between theorists and experimentalists, and solve the difficulties that accompanied the move into the epoch of large-scale, centralized, and cooperative research.

Together these changes transfigured the physicists' approach to research. In the most visible and dramatic fashion the war provided concrete instances of scientific accomplishments, though it remains an open question whether or not the lessons drawn from that experience were actually the ones responsible for the physicists' success. But in a sense that I will develop further below, the major weapons systems – radar, atomic bombs, rockets, and proximity fuses – formed *guiding symbols* that inspired the strategy of much postwar research. Needless to say, one can find earlier instances of one aspect or another of large-scale research: the great philanthropically funded telescopes, Ernest Lawrence's growing array of cyclotrons in the 1930s, and institutes of physics in Europe come immediately to mind – monumental telescopes cost millions of dollars, cyclotrons took several people to operate, and at certain institutes state and scientific concerns shared a common roof. Indeed, at universities such as Stanford and Berkeley, the 1930s saw the establishment of joint endeavors involving physics and electrical engineering. But despite the importance of such successes as the cyclotron and the klystron, before the war there were no physics achievements born of such large physics/engineering efforts that made the continuation of centralized big research seem either inevitable or inarguable. Beginning in the war, however, the physicists' and engineers' large-scale collaborative work on electronmagnetic and nuclear physics-based weapons systems provided just such exemplars. In part as a result of these successes, between 1943 and 1948 key segments of the American physics community came to accept a mutation in the ideal of the physicists' work and workplace. One after another, physics departments began to conceive of a style of orchestrated research that has come to dominate the character of modern investigations in high energy physics, and increasingly in other domains as well.

Expansion and the Repositioning of Physics in the University: Teaching, Surplus, and the Relation of Science to Engineering

Our first need is for a closer analysis of how the expansion affected physics departments, and to achieve this we need to dispose of the myth that changes in scientific planning began only after the guns of World War II had ceased firing. For it was *during* the period 1943–45 that physicists and administrators first debated and set in motion the coming boom. Across the country, from Berkeley to Harvard, pressure to think about postwar expansion began at the top (5).

Harvard

At a meeting on 20 November 1944 the president and Fellows of Harvard College agreed to create a panel whose task was to direct the expansion of Physics, Chemical Physics, and Engineering. Selecting representatives from the various physical sciences at Harvard, President James Conant established a Committee on the Physical Sciences (6). As a driving member of the principal organizations shaping scientific war research, Conant had arrived at a clear conception of the shape of postwar science. It was an image shared by several of his Harvard colleagues, as was evident at that committee's very first assembly. Edwin C. Kemble lobbied for physics in these terms: "*The war* has given a great boost to physics. It has stressed the importance of physics to industry and national defense and has underscored the usefulness of men trained in pure physics when emergency requires that they turn to applications" (7). Consequently, he argued, the university needed to expand to new fields, get the best personnel, and enlarge the group for instructional purposes. In nuclear physics, which was "the [field] of greatest interest," Kemble conveyed the department's desire to keep its existing staff, to add "top caliber" theoretical physicists – for example, Julian Schwinger, Hans Bethe, or Harvey Brooks – and, of course, to augment their total budget with funds for mechanical assistance and construction costs.

To justify the expansion of theoretical physics, Kemble composed a memorandum on 9 December 1944 to the Physical Sciences Panel which began, under the rubric "presuppositions," by contending that there would be a "nation-wide acceleration in the growth of the science of Physics as a result of war emphasis." As Kemble saw it, it was a growth that would take place in two areas. *Solid-state physics* ("the field of properties of matter in bulk") demanded the efforts of a new, more powerful contingent of theoretical physicists, who would be masters of quantum mechanics, statistical mechanics, and chemical thermodynamics. For Kemble, the need for more theory was illustrated by his colleague. P. W. Bridgman, whose superb investigation into the high-pressure domain had nonetheless "undoubtedly fallen short of its maximum potentialities since, to date, he has worked without steady and effective collaboration from theoretical physicists . . ." (8).

Beyond solid-state physics lay *nuclear physics*, which was, by Kemble's lights, "the most spectacular field in physics today." It was there that "the riddle of the physicist's universe is found," amid the cosmic rays, mass spectrographs, cyclotrons, and forms of radioactivity. And as the war was making perfectly clear, medical and chemical applications appeared "manifold," and were accompanied by the even more tantalizing "possibility of unlocking stores of atomic energy [which added] urgent significance to the investigations." Exactly this combination of the purely intellectual and of hoped-for practical consequences characterized the physics community's justification of the needed expansion. In brief, the intellectual argument positioned atomic physics as a stepping-stone to nuclear physics (9).

Speaking for his department, Kemble argued that since World War I the focus of physicists' concern had altered from understanding atomic structure and the structure of simple molecules. Previously, Harvard's own efforts had to a large extent been devoted to spectroscopy of all kinds; but now problems of this type "have largely been solved", and "work in this field operates against a law of diminishing returns," to be replaced by the more alluring problems of nuclear and solid-state physics. Both of these new domains demanded a new, deeper cooperation between experimental and theoretical workers to handle

> the increasingly abstract and complex character of present-day physical theories. This quality is the result of the intensive search for more powerful means of attack on problems of a more and more difficult character. As one consequence of the complexity of the theory, the most brilliant experimental physicist is in need of theoretical collaboration to an extent previously unknown.

At the same time, an increasing number of the "more gifted young men" were choosing theoretical physics. Oppenheimer's presence at the Berkeley Radiation Laboratory was an example for all to see of the theorist's usefulness in joining the skills of experimental and theoretical physicists; the young theorist's contribution "has been of crucial importance in the meteoric rise of that laboratory to its place as principal center for nuclear investigations in this country." Unspoken — but undoubtedly understood — was Oppenheimer's masterful guidance of the Manhattan Project. The Harvard physicists hoped Schwinger or Bethe could enliven theoretical life in Cambridge (10).

Acquiring theorists was, however, only part of a much larger pattern of growth. By the end of December 1944, Kemble was sure enough of the expansion plans to write Conant:

> My personal acquaintance with the state of engineering arts at the time of the last war and at the present time convinces me that it is imperative for our future national safety that the scientific bases of engineering practice shall have far more intensive study than heretofore. We must have a much increased number of analytical engineers with brains and advanced training if we are to hold our own in the technological race (11).

Kemble added that the universities could best aid this cause by prosecuting work in "pure science," which would yield more "in relation to the investment" and attract the best "quality of the men . . . likely to . . . execute it" (12). Here the Harvard physicist embraced the central justifications for expansion that would be repeated over and over again during the following decades: connections between fundamental research and teaching, industrial spin-offs, and military preparedness.

The "modest" investment Kemble had in mind was hiring Julian Schwinger and Edward M. Purcell; adding $25,000 for operating expenses, $30,000 per year for operating the cyclotron, and $30,000 for the physics of metals; a new electronics building at a cost of $100,000; a new mechanical engineering building at $500,000; an electronics research budget of $10,000 per year; further appointments in electronics, mechanical engineering, and aeronautical engineering; construction of a wind tunnel for $75,000; and building an interdisciplinary science center for between 1.5 and 2 million dollars (13). Clearly, the expansion envisioned by plans such as these extend far beyond nuclear or "fundamental" physics and embraced a picture of a much-enlarged program for both physics and engineering. A few days later Kemble contacted his physicist colleagues Jabez Curry Street and Kenneth Bainbridge, requesting them to consider how their respective research areas might participate in the expansion (14).

A boosted physics budget made possible both the new machines and the increased role of theoretical physics. Throughout the American university system the growth of physics was also fueled by a spectacular jump in the number of students. During the war, the armed forces had called upon the physics departments to teach thousands of conscripts the elements of physics so they could cope with a new generation of technical war apparatus, especially radar, radio, rockets, and navigational equipment. After the war the G. I. Bill funded these and other students as they came back to the university in droves. Although providing instruction to so many students often taxed their already depleted resources, wartime instruction also presented universities with an unprecedented opportunity to expand their clientele.

Temporary physics programs had formed during the war at Harvard and Bowdoin in electronics and communication, at MIT in radar, at Los Alamos, and at other institutions as well. At Harvard alone, 5,000 students passed through the pre-radar electronics course. After demobilization the faculty expected that large numbers of veterans from that course would return to further their physics education. As E. L. Chaffee put it on 9 December 1944, the Officer War Training Courses put "Harvard in an advantageous position for attracting students after the war" (15).

Aside from introducing promising students to the university, Chaffee remarked that "the war-training courses have provided us at no expense with a very considerable amount of laboratory equipment. There is insufficient space in the Cruft Laboratory even to store this equipment, say nothing of setting it up for instruction" (16). Instructional equipment could be supplemented by war surplus apparatus suitable for research. Chaffee noted that Harvard's antiradar work and other war projects would offer

> an unusual opportunity to purchase advantageously some very valuable equipment from the OSRD [Office of Scientific Research and Development] projects and perhaps some equipment from the war training courses. . . . We should be prepared to purchase a considerable amount of this equipment. There will [be] available machine tools, obtainable at much reduced prices from the same sources, and I believe we should purchase a considerable amount of this machinery both to increase our present shop facilities and to replace some outmoded and worn machine tools (17).

With this new equipment Chaffee expected that physicists in the postwar period would be able to exploit their newly developed capability to generate microwave signals "by methods which have worked but which are not understood." Such applications included detecting molecular resonances, pulse systems of communication, and high-frequency heating. As Purcell rather overmodestly put it: you didn't have to be too smart to design an experiment with the extraordinary resources offered by the new electronics (18).

Surplus equipment came from many war sources and went to a wide spectrum of users. Connections established between the scientific and defense communities grew, yielding benefits for scientists long after V-J Day. Since they were on contracts from the Office of Naval Research, as late as 1960 physicists at Brookhaven could acquire large armor plating, originally intended for cruisers, to use in neutrino experiments (19). On the West Coast, Robert Hofstadter collected Naval gun mounts on which he could perch his magnetic spectrometer (20). Overseas, A. Gozzini's experiments with surplus pulse-generator circuits and microwave equipment led to his development with Marcello Conversi of the "flash tubes" that played such an important role in cosmic-ray physics and in subsequent work on the spark chamber (21). Phototubes, crucial for scintillation devices, had been much improved and exploited during the war as sources for noise generation in radar countermeasures. As we will see, this type of continuity between wartime and postwar work ran deep; beyond the formation and evolution of administrative organizations such as OSRD, an essential consequence of the war work was the carry-over of physicists' techniques, equipment, and collaborations into the late 1940s.

With the promise of new physicists, new students, and new equipment, departments could embark on major new research projects. Harvard — along with many other universities — was determined to restart its cyclotron program on a much-increased scale. Planning to accelerate both electrons and protons, the nuclear physics planning committee met for the first time in January 1946. They wrote: "From the point of view of physics this program represents a vigorous and progressive plan which should enable Harvard to compete favorably for financial support and, in addition, enhance its attractiveness as a center of research in the nuclear field (22).

When the Harvard physicists turned from war work to accelerator work they brought with them experience from the Manhattan Project (Kenneth Bainbridge), from the Radar Project (Edward M. Purcell, Julian Schwinger), and from Radar Countermeasures (Roger Hickman, John Van Vleck). All served on Harvard's newly established Committee on Nuclear Sciences. After selling their old cyclotron to the government for $200,000, and getting a commitment of $590,000 from Harvard and $425,000 from the Navy, the committee members could begin to plan an 84-inch cyclotron (23). Their physics program included proposals to produce 25 MeV deuterons and 50 MeV alphas in order to explore the nature of the proton, to produce high-energy neutrons, and to extend the wartime fission experiments to elements lighter than uranium and thorium. With accelerated electrons on tap, the planning committee hoped that the cyclotron could also be exploited to pursue radiation therapy, the photodisintegration of nuclei, and the formation of electromagnetic showers (24).

Princeton

At Princeton, as at Harvard, planning for the postwar expansion began long before the euphoric crowds descended on Times Square. On 4 January 1944 a somewhat overoptimistic John Wheeler wrote to H. D. Smyth that he trusted the war would soon be over and he could return to physics shortly. He then went on to formulate a "Proposal for Research on Particle Transformations," of which the "Ultimate Purpose" was "to determine the number of elementary particles, the transformations between them, the combinations which they permit, the nature of their interactions, the relation between these particles and the existing theories of pair formation, electromagnetism, gravitation, quantum mechanics and relativity." As he was witnessing an immensely successful collaboration of theory and experiment at the University of Chicago's Metallurgical Laboratory (Met Lab), culminating in the achievement of fission in December 1942, he clearly saw this as the wave of the future. Wheeler began: "*Plan.* Effective progress calls for the collaboration of experiment and theory. The interaction between the two will be most fruitful, I believe, when in addition both approaches are combined in a single institution, under the same leadership" (25). He argued that such collaboration ought to be the model for physics research at Princeton; the lesson he drew was one taken to heart at institutions across the United States.

Wheeler needed help to solve the main theoretical problems. These included the need to invoke action-at-a-distance theories to eliminate the self-energy difficulty in quantum electrodynamics, the necessity of classifying relativistic quantum field theories, and exploring the theory of positronium. There were also such "associated experimental" problems as nuclear meson capture, mass distribution of cosmic ray particles, and the gamma-ray production of mesons. Like Kemble at Harvard, Wheeler counseled building the theoretical side of the Princeton department, by hiring three theoretical assistants of "the type of Feynman or Jauch," along with some experimentalists of the "type of Luis Alvarez or Bob Wilson." Of course the thirty-three-year-old Wheeler would "[w]elcome collaboration of interested older members of staff," and, again grounding his recommendation on his Chicago experience, he suggested that experimentalists should call on experienced "electronics men" for the design and development of instruments. The program promised, as Wheeler continued to insist after the war, the possibility of sources of energy many times more powerful than all known nuclear reactions, with "obvious implications for the problem of national defense" (26).

As ambitious as young Wheeler's plans may have sounded initially, even before the war's end the Physics Department had begun to address the administration in a new, more confident tone. With the successes of proximity fuses, radar, and then the atomic bombs, physicists — who for years had occupied a decidedly secondary place within the university — acquired a radically improved self-image, evident as they spoke to colleagues and administrators. As the Physics Department put it in a draft of their department report,

> The end of the war finds the department in a praiseworthy but embarrassing condition. The record of the members of the department in war work is laudable, so much so that many of them, particularly in the younger group, are receiving very attractive offers from other institutions and from industry. Such offers are not only attractive in terms of salary but are usually backed by promises of large expenditures for apparatus and equipment. The university must choose between going ahead vigorously, capitalizing the fine record of this department during the war or letting its physicists drift away to such a degree that it may take a generation to restore the department. The first course will require money for men and for equipment, a great deal of money, but it offers a magnificent opportunity, completely in the tradition of the university. We have never been in a better position to push forward in the field of fundamental physical research (27).

Indeed, their position *was* entirely unprecedented.

Physicists everywhere were attracted by the promise of the new technology and science for advancing the physics of nucleons and mesons, but developments in "fundamental" experimental physics at Princeton took on a particular cast. In part this reflected an imaginative style of work that John Wheeler had developed before the war, but the echoes of what he had seen in the Metallurgical Laboratory can clearly be heard in his ideas for postwar research. As early as June 1945 Wheeler had penned a proposal on the future of physics research that gave three goals for the postwar epoch. First, though he voiced doubts about some of their features, he advocated the development of accelerator sources for particles. Among these, he mentioned Luis Alvarez's latest plans to build a linear electron accelerator; Wheeler judged that one would want at least a 5 GeV proton accelerator in order to produce pairs of mesons (28). Second, he wanted Princeton to establish a wide-ranging "ultranucleonics" program. Third — and here he saw the real payoff in physics — he hoped that the department would seize upon cosmic rays as the primary domain in which to search for answers to basic questions, because there and only there could one find the high-energy collisions needed to probe the subnuclear domain. For the cosmic-ray project, Wheeler suggested using Flying Fortresses to hoist experiments and experimenters into the upper atmosphere.

The idea of enlisting bombers for the study of high-altitude cosmic radiation had several appealing aspects. It would alleviate the costs of such exploration for the universities, while leaving control over research apparatus entirely in the hands of the physicists. As for its physics justification, Wheeler noted that only by reaching far into the sky could one study particles with something like 10^{17} electron volts, and therefore exhibit the multiple meson production process that interested him. "This plan calls for army transportation of equipment up to 10 tons to altitudes of the order of 40,000 feet. Research money would in this way be freed for research itself, and for research of a most effective kind" (29).

Finally, Wheeler felt that a survey of the entire field of ultranucleonics was of the highest priority. From what had been learned by cosmic-ray studies, he suspected that it might be possible to transform matter directly and completely into energy on the model of protons being transformed into mesons in the upper atmosphere.

Discovery [of] how to release the untapped energy on a reasonable scale might completely alter our economy and the basis of our military security. For this reason we owe special attention to the branches of ultranucleonics — cosmic ray phenomena, meson physics, field theory, energy production in supernovae, and particle transformation physics — where a single development may produce such far-reaching changes (30).

To reach these dramatic goals, physicists had to make their needs known, and here the survey would play a vital role. It would offer workers in postwar physics "a prospectus of long-range objectives," and it would gain financial support for fundamental physics by making research public and by demonstrating "that scientists in free association can show more vision and judgment on research planning than any centralized government authority." Not least, the ultranucleonic survey would "uncover lines of investigation of evident present or future value to the country's war power" (31). Though these functions were already far beyond the typical prewar involvement of the government in basic research, within a few months Wheeler had in mind new, even more active roles for the government to play in physics.

One such role came from Nazi Germany's development of the dreaded "vengeance weapons," the V-1 "buzz bomb" and the V-2 guided missile. This had been an engineering project of immense scope, costing over $3 billion — fully one and a half times the resources put into the Manhattan Project. The V-2 was brought into the war relatively late, but starting in September 1944 the Germans successfully launched more than 3,000 V-2s, killing almost 10,000 people in England. Soon, however, the Allies began advancing across Europe, and Wernher von Braun retreated from his headquarters at Peenemunde with some 4,000 workers to the V- weapon production facility situated in the concentration-camp complex of Dora-Nordhausen in Thuringia. Installing themselves in the Harz mountains, the V-2 workers successfully evaded the approaching Russian army, eventually surrendering themselves to an American garrison. Under a secret mission code-named "Overcast," the American army shipped the Nazi scientists to the United States to continue missile development work at several sites. The Germans arrived in October 1945, and soon the Peenemunde team was split into groups, working with American industry to produce a variety of rocket types (32).

From 16 April 1946 to 19 September 1952, 64 V-2s were launched from White Sands. The first failed three and a half miles into the air when a fin ripped off and the rocket was destroyed; the next launch, on 10 May 1946, successfully rose to 71 miles (33). For physicists at Princeton, the capture and reinstallation of the German rocket team offered an immediate opportunity. In November 1945 — just a month after von Braun and his associates were brought to White Sands — M. H. Nichols jotted an interoffice memo to Smyth suggesting that the department ought to propose to study optical and electrical phenomena in the upper regions of the earth's atmosphere. At the same time the Princeton group could explore cosmic rays and neutron densities. All this was made possible by "new advances in rocket technique as well as progress here at Princeton and elsewhere in the field of radio telemetering from aircraft and missiles," which would "make possible an extension of present data to regions as high as 500,000 feet" (34).

For some time Wheeler had seen cosmic-ray physics, not accelerator physics, as the primary vehicle for understanding elementary particles. In a memo of January 1946 he stated that "cosmic ray research will take on an even more important role in physics in the next few years," and he advocated immediately setting up a joint experimental and theoretical research group. "Inasmuch as the V-2 firings will not last indefinitely, and inasmuch as the experienced researchers are becoming scarcer every day, it appears that some action [should] be taken as soon as sound decisions can be made." Presumably addressing himself to the department chairman, Wheeler stressed that they would be needing four to six assistants with experience in experimental electronics, nuclear physics, or cosmic rays, as well as experienced cosmic-ray and nuclear physics experimentalists. "Research in physics is starting fresh, and . . . new techniques and new vehicles are now available" — and so it behoved the department to search out consultants from among the best universities, institutes, and weapons laboratories (35).

Wheeler himself headed a Navy-funded project that would handle the telemetric transmission of cosmic-ray data from the German team's missiles. Begun on 1 January 1945 for other purposes, the Navy grant had been extended in March 1946 and was to cover the development of telemetry equipment, while at the same time serving to study cosmic-ray showers and the properties of "mesotrons" through the design, fabrication, and operation of cloud chambers and Geiger counters to be mounted on the V-2s. In July 1946 D. J. Montgomery reported to the Navy that the Princeton V-2 expedition had arrived at White Sands and was making final tests for what they hoped would be the 100-mile-high Princeton Shot on 6 August. At this point $335 thousand had already been allotted, with $250 thousand more to be shared over the next two years between chemistry and physics (36).

Cooperation with the military remained close. Military and elementary-particle problems were interspersed in planning and designing the mission. Both physicists and strategic planners needed a comparison of "Lark" and Naval Research Laboratory telemetering systems, especially with regard to the reliability, intensity of signals, and freedom from disturbances of each system. Both civilians and uniformed personnel had to study radio signal propagation in the ionosphere by transmitting and receiving signals from the missile. In addition, the physicists could use the high-altitude flight to measure cosmic-ray intensity, to distinguish primary cosmic-ray electrons from primary protons, and to measure the neutron productivity in the atmosphere as a function of altitude (37). Such studies directly continued some of the Princeton group's wartime accomplishments in telemetry. In fact, at least one member of the staff, Dr. Walter Roberts, wanted to continue this work on guided missiles at the Johns Hopkins Applied Physics Laboratory. As Wheeler assured his readers, this would ensure a "satisfactory liaison" between the Princeton and weapons laboratories (38).

During the period from 1945 to the early 1950s, the liaison between civilian and military nucleonics functioned well — from both parties' perspectives. The Office of Naval Research (ONR) liberally funded civilian science, and in return the scientists moved easily back and forth between nuclear physics, cosmic rays, and weapons problems. Princeton's nuclear physicist, Milton White, for example, was pleased to report on some recent Princeton instrumentation work that seemed perfectly suited for transfer to the military sector. The laboratory had perfected a new, simple, rugged, and reliable scintillation counter, and White lost no time in alerting ONR to possible defense applications of the new device:

If the U. S. Government has need of α-particle counters, either in connection with plutonium plants, or atomic bombs, there should be set in motion a program for further engineering and quantity manufacture. I can visualize an eventual need of many thousands of counters; if this is correct then our contract with the Navy will already have given the government more than a fair return on the money thus far allocated (39).

White added only that he hoped the Princeton researchers could be spared the engineering details.

Berkeley

On the West Coast, Stanford and Berkeley had no intention of being spared the engineering details. The style of research in the West was somewhat different from that in the East: it was more tightly bound to engineering, and it drew more liberally from philanthropic and industrial sources. Such entrepreneurial physics had brought Berkeley's E. O. Lawrence international fame for his big accelerators paid for from private coffers; as engineering accomplishments big accelerators were unrivaled, though Lawrence was less successful at drawing deep physics from them. As Robert Seidel has so nicely shown, World War II brought a substitution of federal for philanthropic funds, and the important assignment that Lawrence's laboratory direct the electro-magenetic separation of U-235. Lawrence was as well prepared to begin large-scale research as anyone, and shortly was supervising a dramatically bigger laboratory with a wartime expenditure of $692 thousand *per month*. By the middle of 1944 the Radiation Laboratory held a total working population of 1,200 scientists, engineers, and technicians; indeed, once their war work began in earnest the number of engineers at the laboratory never dipped below sixty (40).

After a few months of intense discouragement because of difficulties with his electromagnetic separation facility, Lawrence began escalating his expectations for the postwar period. During the summer of 1944 he began lobbying for ten new isotope separation facilities, leading General Leslie Groves (who was in charge of the Manhattan Project) to some cautious thinking about spending $7—10 million. Just a year later Lawrence began arguing for the rapid expansion of nonweapons facilities, including Luis Alvarez's plans for a linear accelerator and Edwin McMillan's for a synchrotron. Within a few months of the end of the war, Groves authorized $250 thousand in surplus radar sets for the linear accelerator, $203 thousand in surplus capacitors for the synchrotron, $630 thousand for construction in the laboratory, and $1.6 million for six months of operating expenses. Building on earlier experience, engineering and physics had grown together at Berkeley, to make the university one of the models of postwar physical research. In fact, the Berkeley Radiation Laboratory became the pacesetter for the Atomic Energy Commission's development of regional laboratories (41).

Stanford

Stanford, like its Berkeley neighbor, had successfully linked engineering and physics before the war. While Lawrence and his team were building ever-larger cyclotrons, the Stanford physicists were binding electrical engineering to physics as they learned to manipulate microwaves. William Hansen had set the character of that collaboration at Stanford with his stunning development of the "rhumbatron," which set electrons in an oscillatory dance by creating electromagnetic resonances within a copper cavity. Although the device was quickly superseded as a particle accelerator, it formed the core of the klystron, a powerful microwave tube that the Varian brothers designed and deployed in airplane navigation and locating systems. Soon the Sperry Gyroscope Company was underwriting a good deal of the joint physics/electrical-engineering efforts. With the help of their electrical engineer leader, Frederick Terman, Stanford's electrical engineering department built a myriad of radio communications systems around their jewel, the klystron. Gradually, the Stanford engineers transformed the klystron from a fascinating, isolated tube to a standardized component within a whole gamut of microwave circuits (42).

On a technical level, the microwave klystron-based research continued unabated into the early years of World War II: weapon innovations included instrument landing systems and doppler radar. But dramatic changes quickly accompanied the increased pace, scope, and funding of laboratory work. Already in April 1942, Paul Davis (Stanford's general secretary) was writing Terman that "there are many things that could be done under the pressure of the present war situation that will be more difficult to achieve in peace times" –

including ambitious plans for electrical engineering. In August 1942, Stanford issued its "Proposal to Organize the Stanford Resources for Public Service," focusing on how to organize "a vastly augmented program of service on a contractual basis" (43).

After listing suggested projects (from surveys of mineral and industrial resources to the creation of psychological warfare tunes, such as "Marching Civilization"), the August proposal turned to the effects of war work on Stanford. A radical increase in contractual research would provide an opportunity to reorganize faculty administration and to improve the physical plant for more effective war and postwar work. Substantial contracts would bring federal war priority, keeping faculty on campus, creating interdisciplinary research, and engaging a new cadre of talented students who would stay on after the war. Moreover, working on contracts would make Stanford students known to public and private agencies, and might contribute to the long-term development of the West. But above all, government-sponsored research would rocket Stanford to a position comparable to Harvard, Chicago, Caltech, the University of California, and Columbia by bringing in "substantial additional income" (44).

Ironically, while Stanford did greatly expand during the war, its great advocate, Frederick Terman, spent the war years on the East Coast as director of the Radio Research Laboratory in Cambridge (RRL), a facility built to produce radar countermeasures. As the new laboratory grew into a powerful organization, Terman became ever more conscious of the models that Harvard, MIT, and RRL itself presented for the postwar situation at Stanford. He was also deeply impressed with many of the administrators from Harvard — especially with his neighbor, William Henry Claflin, Jr., Harvard's treasurer (45). Terman was quite keen for Stanford's general secretary to speak with Claflin, and the Stanford engineer-administrator soon sought and arranged a meeting between Donald Tresidder, Stanford's president, and Claflin. By December 1943, Terman had concluded that

> the years after the war are going to be very important and also *very critical ones* for Stanford. I believe that we will either consolidate our potential strength, and create a foundation for a position in the West somewhat analogous to that of Harvard in the East, or we will drop to a level somewhat similar to that of Dartmouth, a well thought

of institution having about 2 per cent as much influence on national life as Harvard (46).

Terman went on to set out a plan to "lick" Caltech by equaling it in the physical sciences — since "after all they are only a specialized school, and Stanford is a complete university." In part, Terman wanted a "technical institute" that would create a joint identity among scientific and engineering fields. This alliance would aid in attracting and placing students, raising money, and creating an identity based on special Western areas of strength. One such field was the characteristically Western oil industry, which would link geology, heat transfer, and chemical engineering with the radio industry and accompanying research. Moreover, Terman argued, the competition was softening: Caltech had become "smug," leaving "cracks in its armor" by not developing electrical engineering, and Harvard, Yale, Columbia, and Princeton had thus far slighted the applied sciences in favor of natural philosophy and the humanities (47). The codevelopment of electrical engineering and physics was the hallmark of Stanford physics, as it passed from the Microwave Laboratory to the High Energy Physics Laboratory, and eventually to the two-mile-long Stanford Linear Accelerator.

The Continuity of Technique and Discontinuity of Results

On many levels, then, physics began to change *during*, not after, the war at institutions like Harvard, Princeton, Stanford, and Berkeley. Nonetheless, there is a natural tendency among physicists and historians to overlook the continuity between wartime weapons development work and postwar research, and to reach back before the war to points of common peacetime research. The difficulty may stem from an understandable focus solely on *results*, ignoring the techniques and practices of the discipline. Contributing to the physicists' inclination to elide the effects of war on research is the preponderance of theorists among those who have narrated the discipline's history.

It may also be that war/postwar continuities are slighted because after the war the physics community found itself divided over the opportunities and hazards of the links to weapons research. Physicists walked a tightrope, using government funds to build the machines and teams they needed, but at the same time trying to reestablish a domain of work free from the constraints of a too-closely directed and supervised research. The struggle to maintain that independence also contributed to a vision of the history of physics as, in a sense, skipping lightly over the war years.

But whatever the source of this hesitancy in tracing the continuity of wartime to postwar research, we have inherited a broken narrative. Let me illustrate this point by focusing on the work of the long-productive cosmic-ray physicist Bruno Rossi.

Rossi, an important contributor to cosmic-ray physics before the war, to the war effort itself, and subsequently to high-energy physics, offered the following recollection:

> In 1939, a systematic investigation of air showers was initiated by Auger and his collaborators. Their work, still carried out by means of Geiger-Müller counters, produced results of very great significance. However when, in the late 40's, air shower work was resumed, it became clear that, in order to substantially advance and refine these studies, more sophisticated kinds of detectors were needed (48).

If attention is paid only to the specific results of air-shower research, Rossi's comment makes perfect sense. But instead of halting our historical inquiry at that point, let us descend to a "lower level" of analysis — that is, let us focus on the instruments and techniques of the work in question.

Many of the instruments developed after 1945 to detect air showers were fundamentally linked to war work. In Rossi's particular case, this link was abundantly clear since he, with H. Staub, literally wrote the book on the subject. Their volume, *Ionization Chambers and Counters* (1949), was produced for the National Nuclear Energy Series, Manhattan Project Technical Section. It summarized the advances in electronics and detectors that issued from the radar and bomb projects. Starting in July and August of 1943 Staub had directed a Los Alamos team in charge of improving counters, and Rossi led a group to improve electronic techniques. In September 1943, the two groups were merged into a single Experimental Physics Division group P-6, the detector group, under Rossi (49).

Roughly speaking, their task was to design and implement detector systems that could determine the type, energy, and number of particles emerging from a variety of interactions. Their principal mission was to develop ionization counter systems that functioned in four stages: a first device detected the particle by producing a small current; a second amplified the current; a third separated the signal from unwanted noise; and a final instrument counted and recorded the total number of pulses. Physicists from the two big war projects had improved electronic instrumentation in all four areas — detection, amplification, discrimination, and counting.

An ionization counter works as follows: A charged particle passes through a gas that is contained between two parallel plates at different voltages. Along the particle's track it ionizes gas atoms; the electrons wander toward the positive plate, and the ions toward the negative plate. If the field is not too strong, the charge deposited on one of the plates is equal to the number of ions produced. When these charges arrive at the collecting plate, the current that they produce can be amplified; the shape and height of this current can then be used to determine the charge and energy of the incoming particle.

Rossi's immediate postwar contribution to physics involved the development of fast timing circuits for cosmic rays. His work of the late 1940s built directly on the wartime timing circuits that he had used to link ionization chambers in tests of the Los Alamos "Water Boiler" reactor. For that "Rossi Experiment," as it became known, the Italian physicist set a neutron detector to register the presence of a chain reaction inside the reactor. Using a fast coincidence circuit, the experimenter could count the number of other neutrons emitted during a brief period after the start of fission. In this way Rossi and his co-workers determined the period between the emission of prompt neutrons (those simultaneous with the fission event) and the delayed neutrons (50).

By 1947 military authorities had declassified not only Rossi's electronic contributions, but a compendious batch of 270 Los Alamos technical reports. Immediately, journals on instrumentation brimmed with the new information. Even a cursory perusal of the 1947 volume of *Reviews of Scientific Instruments* indicates the depth of interest in the instrumentation that had been developed in the weapons projects.

Consider just one example from each of the four stages of measurement mentioned above. One way to find a neutron's energy was to scatter it from a hydrogen nucleus inside an ionization chamber. The recoiling hydrogen nucleus, since it is charged, ionizes other particles in the gas; these, in turn, cascade toward the negative plate. The pulse is then proportional to the number of ions, which is proportional to the energy of the recoiling proton, which is proportional to the energy of the original fast neutron. A variety of such "proportional counters" issued from the Manhattan Project, including sensitive ones that could measure the energy of neutrons traveling in a particular direction (51). Signals from devices like these could then be analyzed. The simplest possible device only registered a pulse if its height came above a certain set level. In more sophisticated instruments developed at other laboratories (e.g., by the Chalk River group), separate channels were activated by pulses of varying energy. Such "pulse-height analyzers" gave an immediate energy spectrum; they were (and are) essential instruments in postwar nuclear physics (52). Finally, once the pulses emerged from the discriminator they needed to be counted. Here too a great deal of progress was made during the war. One such device that was designed at Los Alamos to be used with a wide variety of detectors was the "Model 200 Pulse Counter," which, like the other devices just described, was first made public in 1947 (53).

Physicists from the Manhattan and Radar projects disseminated their work widely. One, William C. Elmore, prepared a series of Saturday lectures that he delivered at Princeton in the spring of 1947. The Department of Physics mailed nearly 300 copies of the lectures to physicists all over the United States, and many of the Princeton physicists made quick application of the techniques (54); Robert Hofstadter, to offer one example, recalled that his own work on the inorganic scintillation detector was strongly shaped by Elmore's talks. The next year Elmore published an expanded version of his lectures in the journal *Nucleonics* as a four-part article, "Electronics for the Nuclear Physicist." According to the author, the series constituted in part a "commentary on electronic instruments designed at the Los Alamos Scientific Laboratory, and now employed extensively at various university laboratories" (55).

These particular examples are only a sample of the variety of ways in which wartime ionization detectors, fast electronics, discriminators, and scalars were pursued after the war. Experimentalists used their wartime expertise to design devices for experiments with X rays, electrons, positrons, neutrons, protons, gamma rays, and fission fragments. Perhaps more influential than any of these considerations were the microwave techniques that played crucial roles in accelerator technology after the war: wave guides, transmission lines, klystrons, molecular beams. In addition, there were the benefits of the radiation laboratory efforts – better low-noise amplifiers, lock-in amplifiers, microwave oscillators, which profoundly shaped nuclear magnetic resonance techniques, radio astronomy, and microwave spectroscopy. It is not possible here to speak of the other war-bred technologies that led to calculating machines, computers, and many aspects of programming. Suffice it to say that wartime research had transformed the material culture of the physical sciences.

Collaboration, Work Organization and the Definition of Research

Thus far our attention has been on skills and the instruments of physical research. But the war left another legacy, one not captured in the new research apparatus, or even in the surplus war material that formed such an important basis for experimental work. The war provided a lesson about the nature of research that left an indelible stamp on the physicists who participated in the massive programs at the Chicago Met Lab, the MIT Radiation Laboratory, Berkeley, Oak Ridge, Hanford, and Los Alamos. That lesson concerned large-scale research organized upon complex managerial lines. So it was that just a few weeks after D-Day, Henry Smyth sketched at Princeton a proposal for a new kind of physics laboratory, one that could duplicate the scientific/engineering successes already in hand from the various wartime enterprises.

Smyth titled his July 1944 effort "A Proposal for a Cooperative Laboratory of Experimental Science," and the document reflected on the vast changes facing physics: "The war," he wrote,

> has now reached the stage where it is desirable to make plans for the postwar period and the period of transition. The complete disruption of the normal activities of the

universities and, in particular, of the scientific groups in the universities leaves the whole condition of science in this country highly fluid.

Smyth's remark strikes at a central, and often ignored point: change was facilitated to a large extent because the traditional structures of research, leave-time, personnel, teaching, and interdepartmental boundaries had been radically altered. While "normal procedures" were suspended, deeper and faster mutations could be imposed on the system than would have been possible in peacetime. As Smyth noted, the direction of those mutations would shape the very definition of a physicist, and of physical research:

Forty years ago the physicist working on a research problem usually was largely self-sufficient. He had available a certain number of relatively cheap instruments and materials which he was able to assemble himself into an apparatus which he could operate alone. He then accumulated data and interpreted and published them by himself. Most of the special apparatus that he needed he himself constructed with his own hands. He was at once machinist, glassblower, electrician, theoretical physicist, and author. He instructed his students in the various techniques of mind and hand that were required, suggested a problem, and then let the student work in the same fashion under his general supervision.

But even before the war, Smyth pointed out, physics had been growing more complex. Large laboratories had begun adding specialized technicians to their staffs, including glassblowers and machinists. Even graduate students came to depend on these technicians. Devices such as grating spectrographs dwarfed in size and complexity the simple table-top devices that previously had been sufficient. Thus it was that "a certain amount of cooperation in the use of such installations had to be worked out. But even twenty years ago research problems were largely individual." Only between 1930 and 1945 had this predominantly individual research dwindled, as equipment grew larger, more expensive, and sufficiently hard to handle that it came to require a team of workers to run. Of all such devices, the cyclotron stood out as the most dramatic example, costing in some cases more than the prewar physics budget of fifty laboratories (56).

For Smyth such developments held dangers, as well as promises. If every university aspired to build cyclotrons or betatrons the costs could prove overwhelming. There was a danger that only a few elite institutions would be left in command of research and that consequently the "background of strength in science which [had] grown up so successfully in the country in the past twenty years" would be weakened. Only co-operative research, he felt, could salvage the situation by consolidating various universities and "other institutions" into centralized enterprises (57). In a February 1945 revision of his document, Smyth added that such big projects should not "oppress the individual scientists. Such installations must be the servants, not the masters of the research man" (58).

The OSRD experience of physicists carried, Smyth argued, three lessons: First, the importance of "fundamental science" in solving problems that mere "specialists" could not handle. "The moral which is to be drawn from this experience is that the ultimate technological strength of the country, even for military purposes, rests on men trained in fundamental science and active in research on fundamental problems of science" (59). Second, the war demonstrated the profits that could accrue from "large cooperative research enterprises," cooperation that would extend not only to other scientific fields such as chemistry, biology, and medicine, but to the deep links between physicists and engineers (60). The latter was an alliance with roots dating from before the war, but which bore fruit only in the wartime efforts. Although for security reasons Smyth passed over it, the obvious reference of this section of his proposal is to the Manhattan Project; at the time the revised proposal was written, in February 1945, Los Alamos was only a few months from detonating its first nuclear weapon.

The laboratory of Smyth's dreams clearly echoed a reality that lay, still secret, in the New Mexican desert. He figured 300 square feet per physicist, about 100 physicists, and about 15 foot ceilings, along with 5,000 square feet for large installations. This yielded 525,000 cubic feet, which at \$0.70 per cubic foot would have totaled \$367,500. Industrial production provided the architectural prototype: "The laboratory should be essentially of factory-type construction, capable of expansion and alteration. Partitions should be nonstructural." And with a democratic flourish Smyth appended his intention that "panelled offices for the director or any one else should be avoided" (61).

Soon Smyth found his thoughts echoed in his colleagues' memoranda. One physicist was

genuinely concerned that in the Atlantic coast region we have some possibilities in this field [of cooperative nuclear research] and that all of the government support is not thrown to those other sites which have a prior claim, of course, because of existing facilities (62).

Wheeler also reacted with enthusiasm, in December 1945, to the idea of a multi-university collaborative enterprise, and had no doubt that universities were owed support by the federal government:

Anyone familiar with work on nuclear physics and its applications to military and peace time uses is aware that progress in this field in the United States has now dropped to a very low rate. Scientists are leaving to go to laboratories where they can have conditions of freedom appropriate for independent investigations. The country is losing out because it hasn't been able to work out a system suitable to enlist the participation of the scientists. In addition to this problem of applying science in the country's service, there is also the problem of what the country can do to replenish the scientific capital on which it drew so heavily during the war. The universities paid in years of peace for the fundamental research of which the government took advantage in time of war. The universities need and can rightly call for government support in the future (63).

Such support should come in the form of engineering assistance, and equipment, Wheeler argued. He preferred a system in which the government financed and ran a facility where university researchers could bring their cloud chambers or magnetic spectrographs, make some measurements, and return to their home institutions. A self-administering center would therefore not burden academics more than necessary. For support Wheeler looked to the men and institutions that had built such laboratories in the past: the Manhattan Engineer District, Vannevar Bush, and industrial concerns.

Over the next half year the Princeton physicists combined forces with others in the Northeast to draft a proposal for a nucleonics laboratory that would cost about \$2.5 million (it soon increased to \$15 million, then to \$22 million, and finally to \$25 million). Blessed with support from General Leslie Groves and the Manhattan Project, the planning staff of the budding Brookhaven National Laboratory recruited the building and management expertise of Hydrocarbon Research Inc. In addition, and to the consternation of Oak Ridge, the planners explicitly resolved to crib experience and information from the proven facilities at Oak Ridge (64).

Unlike reactors, which were obviously too big for most educational institutions, cyclotrons hovered for the next few years at the boundary between being too big for universities and too small to merit their own cooperative laboratories. Indeed, Brookhaven's initial attempts to get its accelerator division funded were unsuccessful. Later, when synchro-cyclotrons appeared, the Brookhaven reactor laboratory served as a prototype for collective research at the accelerator, and the model was soon extended elsewhere in the United States and then to Europe as a template for CERN. War laboratories thus clearly provided the managerial models, the technical expertise, and even the personnel for the establishment of postwar collaborative laboratory work.

Concern about the impact of a big cyclotron on university physics is visible in the case of Princeton. Milton G. White reported to Smyth in December 1945 that many members of the department were keen to find a place for themselves at a cyclotron facility, but remained a bit apprehensive about the nature of the research that awaited them: "Dicke is leaning heavily toward elementary particle physics, but not too anxious to press for high energy if the engineering must come out of his hide. He wants to help get the cyclotron under way and then work on some simple interactions." Therefore, to get the accelerator program "back on its feet," White wanted "*very* much to acquire someone who would attend to moving, wiring, redesign problems." In addition, he wanted one of those sought-after types. "a Los Alamos man" (65).

More generally, White foresaw the need to create new positions for the changed environment of the large-scale laboratory, ones outside the traditional academic hierarchy from assistant professor to tenured full professor. "On the one hand," White reported,

> we find physics research going in the direction of complex equipment requiring a supporting staff of highly competent, broadly trained physicists, engineers, chemists and administrative personnel; while on the other hand we have the customary university policy of regarding all scientific employees as likely candidates for academic positions.

Instead of hunting for the "well-rounded" man appropriate to academia,

White advocated a Division of Research that would hire with soft money provided in part by endowment, but significantly supplemented by industrial and government funds (66).

White was concerned about on-campus accelerator physics, whatever might come of proposals for the cooperative laboratory. And expected costs for the cyclotron clearly were going to be high: from \$100,000 to \$300,000. Size would also be significant, since forecasts called for at least 75,000 square feet. Using Smyth's estimate quoted earlier, a plant of this dimension would run at least another \$825,000. In all, White forecast expenses of around \$100,000 per year for the next five years, and even this sum was exclusive of the building and power requirements of the accelerator.

> No crystal ball is required to outline the trend in high energy physics — the trend is up! In not more than six months it should be possible to settle on the part we wish to play in high energy physics, and having settled this we must pick some one accelerator scheme and back it for all we are worth (67).

Faced with such skyrocketing costs, White simultaneously advocated a cooperative nuclear physics laboratory, with funding in large part to be provided by private industry — citing, for example, the Monsanto Company (68). Unfortunately for White's plans, Monsanto declined.

Stanford, by contrast, had already been successful at linking physics, engineering, and private investment. As Stuart Leslie and Bruce Hevly have shown, such scientific entrepreneurship had begun several years before the war (69). Building on Stanford's engineering/physics of the 1930s, in November 1942 William Hansen began advocating the establishment of a "Stanford Microwave Laboratory" to be headed by Karl Spangenberg, an electronics specialist, and drawing on the consulting expertise of H. Skilling and Felix Bloch. When peace came, Hansen wrote the physics department chairman, Paul Kirkpatrick, the microwave lab would draw further staff and equipment from Physics and Electrical Engineering. "At this point, this laboratory goes out and gets a government contract for some microwave job. There should be no difficulty in doing this. With this job will come a priority. . . . Then you start spending money and also, of course, doing the job." Hansen hoped to establish as many attractive fellowships as could be filled, and then to order "equipment — of a sort that will be useful after the war." This should include "machine tools, measuring equipment, books, and any other things that can be used to generate apparatus or research." Played well, the plan would guarantee the "even if we don't have a dime after the war, good physics can be done . . ." (70).

Almost exactly a year later, in November 1943, Hansen elaborated on his initial scheme. It had already been "obvious" before the war that physicists would play a crucial role in industry and that radio engineering (including electronics) was in for rapid expansion. But,

while these trends were obvious before the war, the war has both accelerated and called attention to them. This is especially easy to noticc . . . in the field of radar. . . . The result will be that, after the war, all major universities will be forced to offer strong instruction in these two branches of science.

Precisely because it could strengthen and link the two domains, the microwave laboratory would prove essential. Modeled on the Berkeley Radiation Laboratory, it would remain under the control of the physics department, although the director would control the budget. Indeed, by establishing the laboratory with university funds, the inevitable private support would not dictate the direction of work (71).

Thus by the war's end in August 1945, the microwave laboratory had been in gestation for nearly three years. And with the atomic bomb project passing from top secret to a national obsession, the physics department scaled up its requests. Now, in October 1945, discussions of physicists and physics took on a new, assertive tone, unheard before the war. Of Norris Bradbury, whom Kirkpatrick wanted to lure back to Stanford: "He is thus the head of the group that changed all human history, a continuing group whose power to effect such changes is by no means exhausted" (72). Even Bradbury, it had become clear, had to struggle to keep the staff together at Los Alamos — and the salaries across the country had simply skyrocketed. As an indication of the precipitous rise, Stanford offered one physicist a salary of $3,750, which was met immediately with a $6,000 counteroffer from the University of Chicago. In fact, of a sample of fourteen physicists (almost all between 31 and 41), the *average* salary was $8,460, with a high of $15,000 and a low of $6,000. "Whether one likes it or not [this salary level] reflects the present pronounced bull market in physicists, which has naturally resulted from a short supply and a heavy demand" (73).

Now the Physics Department could address a memo to the president of the university that bluntly asserted in its first paragraph: "It appears to be the manifest destiny of this department to expand." Unanimously the Stanford physicists brought forth several "considerations": (1) ROTC and NROTC students were going to be taking physics courses, as were soldiers in the Army's officers training program; (2) the Consulting Committee on Undergraduate Studies was considering requiring new physics courses; (3) the federal government was probably going to subsidize science students, and "physics will be the first science affected"; and (4) the microwave program was going to draw students, and indeed the whole "war record of physics is causing students to plan careers in this science." Whereas before the war about one freshman per year indicated he wanted to major in physics, the figure now stood at eleven, and would surely rise further with the influx of veterans. On the basis of these facts, the department welcomed "a chance to enlarge the department" from six, to eight or nine on its permanent staff (74).

Enlargement by 50 percent would cost money, and as Hansen had advocated during the war, the source was government-contracted research. Clearly this benefited the universities, but how so the government? Colonel O. C. Maier of the Air Technical Service command put it concisely to Stanford's president, in January 1946. The Army Air Forces, Maier wrote, had two purposes in continuing close cooperation between universities and the military: "We would not only get a good deal of our work accomplished by capable personnel, but in addition build up a pool of trained engineers and scientists, who could be of assistance in War Department research in case of emergency." The choice of tasks spanned the gamut of microwave research: propagation in low, high, ultra-high, very-high, and microwave frequencies; modulation systems, including accurate timing for radar applications; pulse modulators; broadbanding of antennas and circuit elements; magnetrons and klystrons; research on millimeter waves; three-dimensional radar data presentation, and beacon communication; moving-target indicator research; random polarization jammers; flight computers; navigation systems; loran research; and novel radar systems (75). And this was just one prospective offer from one branch of the armed services — supplemented significantly by the Atomic Energy Commission, which soon took over contracting from the Manhattan Project. Within just a few years, workers at Stanford would be moving easily between classified, applied research and the open domain of academic studies.

Conclusion: War and the Culture of Physics

For years we have treated the history of physics as if it simply stopped between 1939 and 1945; only the movement of refugee scientists, bomb building, and the federal administration of a dramatically larger science budget have commanded attention. But if we are going to understand the deeper implications that the conflict had for modern physics and, by extension, for all of the modern sciences, we must look to the techniques and practices of the discipline. In this paper I have followed five lines of continuity: the transfer of technology, the transfer of support, the realignment of physics and engineering, the new relation between theoretical and experimental physics, and the reorganization of the scientific workplace.

Technological transfer consisted, in part, in the invention of new devices: novel accelerator technology, such as the klystron and strong focusing. The new electronic technology of counters, timers, amplifiers, and pulse-height analyzers (among others) all contributed to the postwar burgeoning of the physics of nuclei and particles. But beyond pure invention, the war increased the industrial production of high-performance components. In turn, this capacity made tools available to the experimenter that had previously necessitated custom manufacture. Finally, there was technological transfer at the most literal level — great storehouses of equipment, hundreds of millions of dollars' worth of machinery, that the federal government shipped directly from war-designated activities to the civilian sector as surplus. From the lone researcher picking up a microwave generator in Europe to the best universities of the United States, this infusion of tools and machining equipment transformed the scope and capacities of postwar research.

The financial support for the discipline of physics had also completely altered. The Manhattan Project continued to underwrite many activities, and when it finally closed shop, its sponsorship was quickly taken over by OSRD, ONR, and then by the AEC. Most importantly, at each of the universities discussed here (Princeton, Harvard, Berkeley, and Stanford) it is strikingly clear that the war had trained academic physicists to think about their research on a new scale, invoking a new organizational model. Not only did physicists envision larger experiments than ever before, they now saw themselves as *entitled* to continue the contractual research that both they and the government had seen function successfully during the war. This way of thinking molded both the continuation of wartime projects and the planning for new accelerators and national laboratories. Across the country, with occasional strong dissent, and with different emphases in different regions, the physics community strove to reenact the trilateral collaboration among government, university, and private enterprise.

The expanded institutional base of research permitted more complex relations between theorists and experimentalists, and between physics and engineering. These collaborative relations were firmly established at the huge project centers for scientific warfare: in the Metallurgical Laboratory of Chicago, in the vastly augmented Berkeley Radiation Laboratory, at Harvard's Radio Research Laboratory, in the rocket plants of Caltech, at the MIT Rad Lab, at Oak Ridge, and at Hanford. So it was that when the war ended, it was altogether natural for a Brookhaven, a Stanford Microwave Laboratory, or a rejuvenated Berkeley Radiation Laboratory to assume — from the start — a style of physics that elevated the role of theorists in the shaping of research programs, while keeping scientific engineering front and central. Through these new alliances, the great particle physics laboratories could be built, and the new relationship with the military justified.

Finally, and perhaps most importantly, the war changed physicists' mode of work — in the process redefining what it meant to be a physicist. As Smyth put it so eloquently, forty years earlier the physicist had been at once machinist, glassblower, electrician, theoretical physicist, and author. Just a few years after the war all that had changed: consolidating a prewar trend, the new breed of high-energy physicists were no longer taught to be both theorists and experimentalists — they chose one path or the other. In the place of physicist-craftsmen arose a collaborative association among theoretical and experimental physicists and engineers of accelerator, structural, and electrical systems.

One consequence of these interrelated transformations was a marked shift in rhetoric. The new, often triumphalist tone contained elements of pride in the physicists' contribution; it also signaled a defensiveness, as the scientists struggled to legitimate government funding while avoiding tight restrictions on the prosecution and dissemination of their research.

But the changes in the material culture, organization, and goals of physics went far beyond new turns of phrase. Collectively, these various factors gave rise to a new style of research in nuclear and particle physics. Schematically, it is useful to think of this "industrialization" of university accelerators and national laboratories as having occurred in two stages. The first stage, discussed here, involved an upheaval in the laboratory environment in which nuclear physicists worked. This transformation of the "outer laboratory" began with Lawrence's prewar forays into big science, but became the norm of nuclear physics only in the years 1943—48. In the following decade, the scale change of physics would reach even further into the conduct of experimental high-energy physics. With Alvarez's massive hydrogen bubble chambers, the "inner laboratory" itself, the micro-environment of the experimentalist's measuring and calculating devices, grew, like the outer laboratory, to industrial size (76).

When Henry Smyth solicited a laboratory with "factory-type construction" using "non-structural partitions," and a director's office "without paneling," he was advancing a straightforward architectural request. But in those plans were other, less visible architectures. From the war, physicists had inherited a new sense of mission-directed, team-executed research that required a new human architecture as well — specialization, and collaborations with well-defined leaders (as evident from remarks like Wheeler's, as he speculated on postwar physics). These directors were to lead interdisciplinary nuclear physics programs that, with their movable partitions, could shift priorities as new instruments or questions arose. Gone were the days when Palmer Hall at Princeton or Jefferson Laboratory at Harvard could devote small rooms purely to acoustical or magnetic research. And of course the new research would find its natural place in the lavishly-funded regional laboratories where — as in the war projects — university, industry, and government would work as a triumverate.

At the same time, one senses in the postwar architectural plans an apprehension about the new physics — a sense that the leaders should not isolate themselves or stifle the working physicist. From many of the postwar physics planners one feels a deep tension, socially and intellectually, between the power available through collaboration and the ideological commitment to individual research. It is a tension only incompletely resolved through the Los Alamos model of large-scale, hierarchical teamwork where the physicist could argue with those up the line. Even the director of the huge National Accelerator Laboratory would later write an autobiographical essay entitled "My Fight against Team Research" (77). When Smyth sat down to plan a regional laboratory he assumed there had to be a director's office — but without paneling.

Plans for physics after the war constituted more than a shift in research priorities; they were *simultaneously* reflective of the wartime projects and *determinative* of the future direction of big physics. Whether physicists turned for guidance to the Met Lab or to the Rad Lab they were constructing a new culture by supplanting the guiding symbols of research. No longer could the image of laboratory work come from the precision interferometry studies of an Albert Michelson. Now, the cultural symbol of physics would originate in a Los Alamos, a Brookhaven, or a National Accelerator Laboratory.

In speaking of the guiding role of cultural symbols, I have in mind something similar to the role Clifford Geertz accords symbols — agreed-upon programs for future action, not mere emblems such as flags (78). It is in this more robust sense of the term that the weapons projects were symbols. In the context of America in the mid-1940s, the Manhattan Project was far more than an indicator of the usefulness of physics; it was seen as a prescription for the orchestration of research. As a representation of how technical, physical, military, and political activities could coalesce, the wartime laboratory became the site for a mutation in the culture of physics.

After the fact, it is hard to grasp how abruptly physics was transformed from one among many university activities, all roughly on a par, to a massive enterprise that was consulted on subjects from university decisions to foreign policy. Suddenly, physics had both wealth and power. And physicists looked to their wartime experience not only to legitimize continued funding, but also for the administrative and work relationships that would govern them and succeeding generations of scientists. As a whole, the physics community had constructed a new identity for itself in the turbulent years between war and peace (79).

Acknowledgments

Citation of manuscript sources is by permission of Harvard University Archives, the Princeton University Archives, and the Stanford University Archives. The staffs of these libraries were immensely helpful, and I gratefully acknowledge their assistance. For comments and suggestions I am indebted to B. Hevly, R. Hofstadter, C. A. Jones, D. Kevles, R. Lowen, A. Needell, S. S. Schweber, and J. A. Wheeler. This work was conducted with support by the National Science Foundation, SES 85-11076, and the Presidential Young Investigator Award.

Notes

1. Galileo, *Dialogue Concerning the Two Chief World Systems*, transl. Stillman Drake (Berkeley and Los Angeles: University of California Press, 1967), pp. 186—187.
2. See, for example, Einstein's popularization of relativity first published in 1916: *Relativity: The Special and General Theory* (New York: Crown Publishers, 1961).
3. R. P. Feynman, "The Theory of Positrons," *Physical Review* **76** (1949), 749.
4. On the World War II scientific organizations and their effects on research see Irvin Stewart, *Organizing Scientific Research for War* (Boston: Little, Brown and Company, 1948); D. Kevles, *The Physicists* (New York: Alfred A. Knopf, 1978); B. Hevly, "Basic Research within a Military Context: The Naval Research Laboratory and the Foundations of Extreme Ultraviolet and X-Ray Astronomy 1923—1960" (Ph.D. dissertation, Johns Hopkins University, 1987), esp. chap. 2; and A. Hunter Dupree, "The *Great Instauration* of 1940: The Organization of Scientific Research for War," in G. Holton (ed.), *The Twentieth-Century Sciences* (New York: Norton, 1972), pp. 443—467. On the postwar contributions of one important agency see S. S. Schweber's contribution to this volume, "The Mutual Embrace of Science and the Military: ONR and the Growth of Physics in the United States after World War II."

5. On the postwar work at Berkeley see R. Seidel, "Accelerating Science," *Historical Studies in the Physical Sciences* **13** (1983), 375—400.
6. J. Conant to Kemble, 28 November 1944; summary of meeting of 20 November 1944, in Hickmann Papers, Physics Department Historical Records, box 3, ca. 1930—65, Harvard University Archives (hereinafter abbreviated as HP). Already in the summer of 1943, Conant was deeply involved in postwar planning for military/scientific relations, though the early speculations did not lead directly to accepted policy. See the excellent book on the military's expectations for their postwar condition: M. Sherry, *Preparing for the Next War: American Plans for Postwar Defense 1941—45* (New Haven and London: Yale University Press, 1967), pp. 137—138.
7. File cards in handwriting of E. C. Kemble, HP (6).
8. Kemble, "Panel: Physical Sciences. Memorandum on Proposals for the Development of the Department of Physics," 9 December 1944, HP (6).
9. *Ibid.*
10. *Ibid.*
11. Kemble to Conant, 22 December 1944, HP (6).
12. *Ibid.*
13. Kemble, "Tentative Summary of Proposals before Physical Sciences Panel," HP (6).
14. Kemble to Street, 30 November 1944; Kemble to Bainbridge, 30 November 1944, HP (6).
15. Chaffee, "Expansion of Research and Instruction in the Cruft Laboratory," HP (6). We still have relatively few systematic data on the effects of this teaching: were there lasting effects on the physics curriculum? Where did the students go after the war: to academic pursuits? industrial positions? military assignments?
16. *Ibid.*
17. *Ibid.*, p. 8.
18. E. M. Purcell, interview 26 May 1987. Purcell discovered the 21-centimeter line, and won the Nobel Prize for his co-invention of Nuclear Magnetic Resonance — both exploited Rad Lab techniques.
19. M. Schwartz, interview 20 October 1983.
20. R. Hofstadter, H. R. Fechter, and R. H. Helm to Commandant, Mare Island Naval Shipyard, 9 February 1952, Hofstadter private papers, Stanford University.
21. See P. Galison, "Bubbles, Sparks, and the Postwar Laboratory," in L. Brown, M. Dresden, and L. Hoddeson (eds.), *From Pions to Quarks: Elementary Particle Physics in the 1950's* (Cambridge: Cambridge University Press, forthcoming).
22. "Proposal for Nuclear Physics Program at Harvard," submitted 2 February 1946 and enclosed with minutes of the first (11 January 1946) meeting of the Committee on Nuclear Physics [later Committee on Nuclear Sciences], in file, "Index and Records," box "Committee on Research and Nuclear Sciences, Records 1946—1951," Harvard University Archives.
23. P. H. Buck to R. Hickman, 5 October 1946, file, "Miscellaneous Financial Material and Correspondence," in box "Committee on Research and Nuclear Sciences, Records 1946—1951," Harvard University Archives.
24. "Proposal for Nuclear Physics Program at Harvard" (22).

25. J. A. Wheeler to H. D. Smyth, 4 January 1944, file "Postwar Research," in Physics Department Departmental Records, Chairman, 1934–35, 1945–46, no. 1, Princeton University Archives (hereinafter, PUA).
26. *Ibid.*
27. "Department of Physics Report to the President 1944–45," Physics Department Chairman's Correspondence 1942–43, 1943–44, no. 16, PUA.
28. J. A. Wheeler, "Three Proposals for the Promotion of Ultranucleonic Research #6: H. D. S.," 15 June 1945, copy to Smyth, in Physics Departmental Records, Chairman 1934–35, 1945–46, no. 1, PUA.
29. *Ibid.*
30. *Ibid.*; and see J. A. Wheeler, "Elementary Particle Physics," *American Scientist* **35** (1947), 177–193, 223, 170, 172, 174.
31. Wheeler, "Three Proposals" (28).
32. F. Ordway III and M. R. Sharpe, *The Rocket Team. From the V-2 to the Saturn Moon Rocket* (Cambridge, Mass: MIT Press, 1982); see also the important article by Linda Hunt, "U.S. Coverup of Nazi Scientists," *Bulletin of the Atomic Scientists*, April 1985, 16–24.
33. Ordway and Sharpe, *op. cit.*, 1982 (32), pp. 353–354.
34. M. H. Nichols to H. D. Smyth, 26 November 1945, file, "Postwar Research," Physics Department Records, Chairman 1934–35, 1945–46, no. 1, PUA.
35. No author listed [probably Wheeler], "Program in Cosmic Rays," January 1946, file "Postwar Research," Physics Department Departmental Records, Chairman, 1934–35, 1945–46, no. 1, PUA.
36. D. J. Montgomery, "Annual Report on Project Assisted by Outside Funds," 23 July 1946, file, "A-475 Wheeler," Laboratory and Research Files, 1929–54, box I of 5, PUA (see also "Elementary Particle Projects as of 6 May 1946," in same file). Among physics goals listed in this report were: determination of total cosmic-ray intensity, meson production, neutron intensity, multiply charged particles at rocket altitudes; study of radio propagation in ionosphere; telemetry tests; pressure/temperature studies; and coordination with the Schein group's ground cloud-chamber and balloon tests.
37. Wheeler, "Appendix IV – General Survey of the Princeton Project Program – Cosmic Rays and Telemetering," 28 August 1946, file "A-475 Wheeler," Laboratory and Research Files, 1929–54, box I of 5, PUA.
38. *Ibid.*, see under "Guided Missile Developmental Work."
39. M. G. White to Urner Liddel, Nuclear Physics Section, ONR, 11 July 1947, file "761," box V of 5, Laboratory and Research Files, 1929–54, PUA.
40. Seidel, *op. cit.*, 1983 (5).
41. *Ibid.* See also R. Seidel, "A Home for Big Science: The Atomic Energy Commission's Laboratory System," *Historical, Studies in the Physical Sciences* **16** (1986), 135–175.
42. On Stanford's early combination of physics and engineering see the excellent article by S. W. Leslie and B. Hevly, "Steeple Building at Stanford: Electrical Engineering, Physics, and Microwave Research," *Proceedings of the IEEE* **73** (1985), 1169–80.
43. P. Davis to F. Terman, 18 April 1942, Terman Papers, SC 160, 1:1:2, Stanford

University Archives (hereinafter SUA); unsigned typescript, "A Proposal to Organize the Stanford Resources for Public Service," 24 August 1942, Terman Papers, SC 160, 1:1:2, SUA.

44. "Proposal to Organize the Stanford Resources" (43).
45. Terman to Davis, 23 August 1943, Terman Papers, SC 160, 1:1:2, SUA.
46. Terman to Davis, 29 December 1943, SC 160, 1:1:2, SUA; partially cited in Leslie and Hevly, *op. cit.*, 1985 (42).
47. *Ibid.*
48. Bruno Rossi, "Development of the Cosmic Ray Techniques," in A. Berthelot (ed.), *Colloque International sur l'Histoire de la Physique des Particules*, in *Journal de Physique*, no. 12, vol. 43 (1982), p. C8—82.
49. D. Hawkins, "Part I: Toward Trinity," in D. Hawkins, E. C. Truslow, and R. C. Smith, *Project Y: The Los Alamos Story* (Los Angeles: Tomash Publishers, 1983), p. 90.
50. *Ibid.*, pp. 104—107.
51. J. H. Coon and R. A. Nobles, "Hydrogen Recoil Proportional Counter for Neutron Detection," *Reviews of Scientific Instruments* **18** (1947), 44—47.
52. Before the war there were essentially two ways to obtain an energy distribution. One could record the pulses photographically with an oscillograph, which was cumbersome and required large amounts of film. Or one could employ a counting circuit with a discriminator that would record the number of counts N above an energy amplitude E; the resulting "bias curve" (N versus E) then had to be reduced after the experiment by taking $N(E) = dN/dE$, and plotting this quantity against E. See H. F. Freundlich, E. P. Hincks, and W. J. Ozeroff, "A Pulse Analyser for Nuclear Research," *Reviews of Scientific Instruments* **18** (1947), 90—100. As of February 1947, descriptions of the other devices still had not been published; e.g., E. A. Sayle, British Project Report, January 1944 (cited in Freundlich et al.).
53. W. A. Higinbotham, J. Gallagher, and M. Sands, "The Model 200 Pulse Counter," *Reviews of Scientific Instrument* **18** (1947), 706—714.
54. Committee on Project Research and Inventions, Princeton University, "Proposal for Continuation of Research Project in Proton-Nuclear Reactions for the Year 1948—49," 5 March 1948, loose in box V of 5, Princeton Physics Department Laboratory and Research Files, 1929—54, PUA.
55. W. C. Elmore, "Electronics for the Nuclear Physicist," Parts I—IV, *Nucleonics* **2** (nos. 2—4, 1948), 4—17, 16—36, 43-55, 50—58; quotation on p. 4 of Part I. Together Elmore and Matthew Sands wrote a book, *Electronics: Experimental Techniques* (New York: McGraw-Hill, 1949), that was widely distributed and translated into several languages.
56. H. D. Smyth, "A Proposal for a Cooperative Laboratory of Experimental Science," 25 July 1944, file "Postwar Research 1945—46," Physics Department Departmental Records, Chairman, 1934—35, 1945—46, box 1, PUA.
57. *Ibid.*
58. H. D. Smyth, "A Proposal for a Cooperative Laboratory of Experimental Science," revised version, 7 February 1945, file "Postwar Research 1945—46" (56).
59. Smyth, revised version of "Proposal," p. 3 (58).

60. Smyth, "Proposal," versions of 25 July 1944 and 7 February 1945 (56, 58).
61. Smyth, revised version of "Proposal" (58).
62. William W. Watson to Smyth, 23 June 1945, file "Postwar Research 1945—46" (56).
63. J. A. Wheeler, draft of "Proposal for Cooperative Laboratory," 11 December 1945, file "Postwar Research 1945—46" (56).
64. For my discussion of the origin of BNL I have relied on the excellent article by Allan Needell, "Nuclear Reactors and the Founding of Brookhaven National Laboratory," *Historical Studies in the Physical Sciences* **14** (1983), 93—122.
65. M. G. White to H. D. Smyth, 20 December 1945, file "Postwar Research 1945—46" (56).
66. M. G. White to H. D. Smyth, 6 May 1946, file "Postwar Research 1945—46" (56).
67. White to Smyth, 20 December 1945, file "Postwar Research 1945—46" (56).
68. White to Charles Thomas, Monsanto Chemical Company, 18 September 1945, file "Postwar Research 1945—46" (56).
69. See Leslie and Hevly, *op. cit.*, 1985 (42).
70. Hansen to Kirkpatrick, 6 November 1942, Terman Papers, SC 160, 1:1:7, SUA (43).
71. W. Hansen, "Proposed Micro-Wave Laboratory at Stanford," typescript, 17 November 1943, Terman Papers, SC 160, 1:1:8, SUA.
72. Kirkpatrick to Tresidder, 1 October 1945, Tresidder Papers, SC 158, 1:4, SUA.
73. *Ibid.*
74. Kirkpatrick to Tresidder, 6 November 1945, Tresidder Papers, SC 151, 25:1, SUA.
75. Colonel Oscar C. Maier to Office of the President, Stanford University, 11 January 1946, Tresidder Papers, SC 158, B1:4, SUA.
76. For more on the "inner" and "outer" laboratories and an extended discussion of the transformation of the inner laboratory in the 1950s, see P. Galison, "Bubble Chambers and the Experimental Workplace," in P. Achinstein and O. Hannaway (eds.), *Observation, Experiment and Hypothesis in Modern Physical Science* (Cambridge Mass: MIT Press, 1985), pp. 309—373; and Galison, "Bubbles, Sparks" (21).
77. R. R. Wilson, "My Fight against Team Research," in G. Holton (ed.), *The Twentieth-Century Sciences* (New York: Norton, 1972), pp. 468—479.
78. Clifford Geertz, *The Interpretation of Cultures* (New York: Basic Books, 1973), esp. pp. 44ff.
79. While the organizational features of the World War II weapons projects endured into postwar "pure" science, the extraordinary political consensus that bound the civilian physics community to the defense physics establishment did not. During the 1950s, for many reasons, the two scientific groups began to bifurcate. Some of these reasons were institutional — the slow decline of the General Advisory Committee, the rising capability of weapons laboratories outside universities; some were political — splits over the hydrogen bomb, the ABM system, the role of secrecy, the cold war; and some were physical — high-energy physics decisively split from nuclear physics both theoretically (e.g., current algebra, field theory), and experimentally with the exploitation of devices that were useful in one field

but not in the other (e.g., bubble chambers). This is not to say that the two communities should be seen as completely autonomous — links remained through advisory panels, students, funding sources, and, most of all, shared technologies. But in the decades following the atomic bomb, the nature of the connection between the civilian and military scientific establishments changed from one of joint enterprise to one of shared resources. I will to pursue these issues elsewhere.

THE CRYSTALLIZATION OF A STRATEGIC ALLIANCE: THE AMERICAN PHYSICS ELITE AND THE MILITARY IN THE 1940S

PAUL K. HOCH

University of Nottingham

Central to the rate and direction of progress in American science in the past two generations was the crystallization of what I will call a military-industrial-scientific alliance in the decade of the 1940s, in which a small elite of physical scientists played a leading part. This process was preceded by the development of separate military, industrial, and scientific elites over at least the previous century, and by their gradual coalescence beginning roughly around the First World War. The merging of these elites was facilitated by the growth between the world wars of special intermediate (or boundary) elites who were able to mediate between two or more of the alliance's constituences. One such mediator was Vannevar Bush, who in the early 1930s, as a dean of engineering and vice-president of the Massachusetts Institute of Technology (MIT), came into continuous contact with the major electrical and electronics firms that provided future job destinations for his students, as well as support for the Institute generally. Then, in the second half of that decade, as a member of the National Advisory Committee for Aeronautics (NACA), and later as chairman of NACA and president of the Carnegie Institution in Washington, he came into continuous interaction with others in the foundation and governmental elites who were interested in the technological and military potential of science. Finally, in the years immediately before the Second World War, as a member of the organizing group of the National Defense Research Committee (NDRC) — and subsequently as director of the broader Office of Scientific Research and Development (OSRD) — he assumed a central

E. Mendelsohn, M. R. Smith and P. Weingart (eds.), Science, Technology and the Military, Volume XII, 1988, 87–116.

role in the crystallization in embryo (and in his own person) of such an alliance.

This does not, of course, mean that the terms of the alliance were well established — that there was not, for example, vigorous debate about the relative positions of the constitiuent elites and about the ideological vocabularies of their collaboration (which are always more responsive to some elements of the situation, and to some social forces, than to others). A great deal of debate, both during and after the war, was to center on the professed need to preserve "security" in science — even at the cost of imposing security screening and loyalty investigations on scientists, or of denying them information that (it was said) they did not "need to know." The scientists, on the other hand, argued that the security apparatus and the compartmentalization of information weakened their ability to do science (1) — an activity built on trust, in which it was inherently difficult to say in advance who needed to know what, or which apparently unrelated pieces of information might provide the key to new understandings (2). In the celebrated security cases of J. Robert Oppenheimer and Edward U. Condon, there was vigorous clashes about the latitudes for political belief and association of scientists.

These cases attracted gigantic public controversy, showing the extent to which the Los Alamos generation of scientists had arrived on the political landscape. Inasmuch as political authority has always been augmented by military power, atomic scientists like Oppenheimer — as progenitors and custodians of atomic weaponry — had become of considerable import. Their atomic bombs, reactors, and huge "atom-smasher" machines had become symbols of the system's strength and ideological legitimacy (3). A modern priesthood, they had to be kept safe from any ideological taint, not just so that they would not betray the arcane "secrets" of their esoteric craft, but even more so that their political purity could not be called into question. Science would be mobilized to support the existing order only to the extent that scientists did not themselves question that order. Indeed, the relation between science and the modern state, in both its capitalist and communist incarnations, rests on the premise that scientists confine themselves to performing scientific and technical functions: providing the *means* of power, and leaving its *ends* to political leaders. This is somewhat similar to the idealized relation between political leaders and their military subordinates, who must also refrain from becoming overtly "political". But the two cases are, in fact, very different, in that scientists are not just technicians of power, but also knowledge-seekers and educators — and in these roles they cannot so easily confine themselves either to compartmentalized areas of knowledge or to purely technical and functional roles. Nor by 1944 could the atomic scientists so easily separate their work from the aims, including political aims (and very destructive consequences), to which that work might be applied. This led to more or less constant tension over the next generation between the atomic scientists, the military, and political leaders over the role of science (and of scientists) in the political arena (4). To the extent that the elite managers of scientists shared the world view of the military, industrial, and governmental elites with which they conducted business — especially in regard to the purely functional character of the role allotted to the rank and file — there was also some considerable tension between the scientific elite and ordinary scientists. On the other hand, the scientific elite's struggles with their military counterparts for status and autonomy bear a close resemblance to those of other intellectual and disciplinary elites (5).

1. Origins of the American Scientific Elites

The growth of an American scientific elite in the nineteenth century was closely linked to the "professionalization" of a leading element of scientists, centered in such institutions as universities and the Smithsonian Institution (6). It was in the course of the Civil War that, in 1863, a leading group constituted themselves as the National Academy of Sciences (NAS) and acquired the right to advise the national administration in Washington on matters of science affecting the public welfare. The Civil War also brought about the spectacular growth of firms such as the Du Pont Company, who assumed the role of leading munitions suppliers to the substantially expanded American military (7). In the two generations after the conflict there was a slow, parallel growth of American scientific and industrial institutions, but with comparatively little important interpenetration between them. Interactions between university and industrial science expanded greatly with the rise of industrial research laboratories, especially in the electrical and chemical industries in the first decades of the twentieth century (8). The most spectacular growth and spread of such facilities occurred in all the countries of the Western Entente (including Japan) during the First World War, especially as they all faced the common need to develop their own abilities to synthesize the chemicals and optical supplies previously predominantly imported from Germany.

During the First World War, as in the Civil War, there was again a pronounced move to put the scientific elite on a firmer footing in collaboration with the governmental elite; in this case, by the self-constitution within the National Academy of an elite group — the so-called National Research Council (NRC) — that defined its initial task as providing overall coordination for research (e.g., centered on tracking enemy submarines) in the military emergency. The NRC had close relations with industrial interests, especially at General Electric, and indeed was at first financed by the Engineering Foundation (9). The council subsequently attempted to provide for its continuance beyond the war, obtaining a presidential mandate from Woodrow Wilson to

> stimulate ... and promote cooperation in research; serve as a means of bringing American and foreign investigation into active cooperation with the scientific and technical services of the War and Navy Departments ... ; direct the attention of scientific and technical investigators to the present importance [in 1918] of military and industrial problems in connection with the War, and to aid in the solution of these problems by organizing specific researches . . . (10).

This was in embryo a perfect anticipation of the 1940s alliance, but the social environment was still too weak to support it; in particular, none of the three elites yet commanded sufficient resources, especially after the postwar demobilization.

The NRC's chairman, the astronomer George Ellery Hale, saw the Council's participation in the war effort in part as a means to guarantee the continuation of a postwar role in a national research effort (11). As early as June 1, 1916, according to Hale's diary, he was already contemplating a continuing postwar alliance "for the increase of knowledge and the advancement of the national security . . ." (12). As Ronald Tobey has made clear, the NRC saw its role as one of promoting cooperation "between universities, private research institutions, scientific societies and the government . . . all [having] been brought into contact with industry . . . [which] would be the basis for industrial progress following the war" (13). More broadly, Tobey sees this as part of the drive of the rising scientific elite for a "national" science under their own control, in support of national welfare and security, in a manner similar to that expressed by other "mandarin" ideologies elsewhere (14). The wartime experience of the NRC leaders "led them to create a new basis for science in America" (15), summarized in a document submitted in 1917 ("Suggestions for the International Organization of Science and Research") that advocated that direct contacts between science, industry, and the military be promoted in all the Allied countries after the war. This was achieved in America by securing President Wilson's executive order establishing (but not financing) NRC on a continuing footing, and by a grant of $5 million from the Carnegie Corporation for a National Academy building and a specifically NRC endowment (16). In Britain, similar wartime developments led to the establishment of the Department of Scientific and Industrial Research, with scientific and industrial advisory committees of the Privy Council Office attempting to coordinate research within their specific jurisdictions.

In fact, the NRC's main role in the succeeding decade turned out to be the distribution of the Rockefeller-financed NRC fellowships in physics and chemistry, which did a great deal to build up the research strength of American universities in these areas. Nevertheless, the NRC in the 1920s and 1930s was still in many ways the junior partner in relation to the major philanthropic foundations and the foundation elite – which by then included such former physicists as Max Mason and Warren Weaver at the Rockefeller Foundation, and would see Bush at the Carnegie Institution. Through its General Education Board, the Rockefeller Foundation between 1925 and 1932 contributed $19 million to the expansion of key American university science departments, including, perhaps most notably, those at Princeton, Cal Tech, Berkeley, and the Massachussetts Institute of Technology. The NRC was a long way from being able to match such resources. In the 1930s, F. G. Cottrell's Research Corporation and the Rockefeller Foundation put up the bulk of the finance for the first cyclotrons, and thus contributed to the technical (and ideological) infrastructure of what would later emerge after the war as "big physics." The Rockefeller Foundation also supported for two years, from 1933, President Roosevelt's Science Advisory Board (SAB), which was headed by MIT's president Karl Compton and sought to apply science and technology to the problems of industrial recovery in the economic depression. There was unfortunately continuing tension between the SAC and the National Academy, both about the relative participation of natural and social scientists and, more seriously, about whether government funding constituted an interference in science. By the end of 1935 the SAC had quietly expired, its supposed advisory functions devolving by presidential edict back onto the NAS. The SAC has since been described as another somewhat premature "attempt to create a central scientific organization both for the government and for the country" (17).

Two quasi-governmental bodies did, achieve more success in the 1930s. The National Cancer Institute, formed in 1937 as part of the Public Health Service, was to distribute without significant protest a large number of grants-in-aid to private institutions and would also initiate a number of advanced training programs. (This is part of a more general political phenomenon in America, in which central coordination of the fight against disease by either the central government or the Rockefeller Foundation was accepted much earlier than — and provided a point of entry for — later efforts along the same lines in support of science, education, or industry.) Likewise, the National Advisory Committee for Aeronautics (NACA) — originally formed in World War I — provided a valuable training ground for aspiring scientific/governmental mediators like Bush, who later claimed it as his model for OSRD (18). In the thirties NACA developed close ties to university engineering departments and supported research relevant to its mandate (19). In addition, the National Resources Committee (successor to the National Resources Planning Board) initiated a monumental study of university, industrial, and governmental research, whose title, *Research — A National Resource*, provided a strong incentive to see the different elements of the research community in the universities, government, and industry as parts of a whole that was vital to national welfare.

At the interface between the universities and industry, an intermediate elite was already well developed by the 1930s. A key member was Frank Jewett, president of Bell Laboratories and later, in the decade, also of the National Academy. Another was Karl Compton, former physics professor and departmental chairman at Princeton, and a long term advisor and consultant to General Electric, whose president, Gerard Swope — in his role as chairman of the MIT Corporation — managed to propel Compton into the presidency of the Institute (20). As chairman of the board of the American Institute of Physics (AIP), an umbrella body formed in 1931 to mediate between the various physical societies (and between industrial and university physicists), Compton was also in a position to play an influential role in the alliance between academic and industrial physics. The AIP got its major funding from the Rockefeller Foundation and by subscription from the major corporate research laboratories (21). As previously mentioned, Compton's vice-president at MIT, Vannevar Bush, was also to play an important mediating role. Indeed, it was basically these three men, together with Harvard's President James Conant, who persuaded President Roosevelt in 1940 to form the eight-man NDRC, two of whose members were appointed by the secretaries of War and of the Navy, thus establishing in committee form the various elites who were party to the coming alliance (22).

2. The Development of a Big Physics Alliance

Vannevar Bush sought to create a civilian-run OSRD in 1941 precisely because he feared the effects on science of the atmosphere of military restriction with which he and other scientists had had to contend in the First World War. He argued that only under a civilian leadership could American science be fully mobilized. Likewise, President Roosevelt's order creating the OSRD provided that it was to "serve as a center for mobilization of scientific personnel and resources of the Nation in order to assure [their] maximum utilization . . . [for] defense purpose"

(23). The reactions of the military to the new body ranged from the absolute fury of the now partly superceded Naval Research Laboratory to the apparent indifference of the commanding generals and admirals, who were relatively slow to see just how crucial scientific research could be to modern warfare (24). This alliance has threatened to come apart on many occasions — for example, over the reluctance of the military to admit the scientific elite to its wartime strategy committees, the conditions of security screening and information compartmentation in the Manhattan Project, and the question of whether the Los Alamos Weapons Laboratory was to be a military or a civilian installation (this was nicely fudged by having Oppenheimer as a civilian director, with General Groves in overall command. The tensions continued in the postwar debates about whether there should be military or civilian control over atomic energy (and later, over space research), or whether the military or the proposed National Science Foundation was better suited as a patron for university research.

With the development of the wartime alliance, university physics between 1939 and 1945 was to be transformed from a situation of relatively small staffs and low budgets, with minimal government funding, to one of far larger groups of personnel (especially teams of research personnel), relatively lavish government funding (channeled through the military and other federal agencies), copious equipment and encouragement for research. With the establishment of OSRD, physics left behind its "shoe-string and sealing wax" era and became Big Science. The graph of federal funding for scientific R&D assumes a dramatically upward slope during the Second World War, and — despite some immediate postwar demobilization — it then continues upward, regaining a dramatic momentum particularly during the Korean War. According to the prewar chairman of the University of Rochester's physics department, Lee DuBridge, the main factor in this growth was "the increased prestige [for physics] brought about by the war and postwar developments" (25). The immediately enabling factor for the expansion was, of course, the tremendous growth of military-financed R&D, brought about by the increased prestige (26). Membership in the American Physical Society nearly doubled between 1938 and 1946, from 3,341 to 5,714 members, and then spurted ahead so that it stood at 8,100 by 1949, and at approximately 16,000 by 1960. The Institute of Radio Engineers went from 6,000 members in 1940, to 30,000 in 1950, growing to about 90,000 in 1960; much of its growth derived from the wartime radar program and the 1950s boom in solid-state military electronics (27) (See Table 1).

After the war, the military and the heavily militarily oriented Atomic Energy Commission were to provide something like 90 percent or more of the support for American university physics research, a percentage that was re-allocated but in no way diminished by the creation of the militarily oriented National Aeronautics and Space Administration to succeed NACA in the latter 1950s (see Table 2). During the

TABLE 1
Economic sectors in which government procurement forms more than one-third of final demand (United States, 1958).

Sector	Government purchases (%)
Aircraft and parts	91.4
Primary nonferrous metal manufacture	57.0
Chemical and selected chemical products	52.3
Electronic components and accessories	52.0
Ratio/TV/communications equipment	37.0
Scientific and controlling instruments	35.7

Sources: OECD, *Government and Technical Innovation* (Paris: OECD, 1966); Gary Werskey, "Science and War in the Twentieth Century," in *War in Our Own Day* (Milton Keynes: Open University Press, 1973), p. 34.

TABLE 2
Percentage shares of military and militarily related R&D expenditures as a proportion of total government R&D (1960–61).

Country	Defense	Space	Nuclear	Total
United States	68.1	9.1	10.7	88.5
United Kingdom	64.5	0.5	14.7	79.7
France	41.5	—	27.5	69.0

Sources: Chris Freeman et al., "The Goals of R&D in the 1970s," *Science Studies* **1** (1971), 357–406; Gary Werskey, "Science and War in the Twentieth Century," in *War in Our Own Day* (Milton Keynes: Open University Press, 1973), p. 32.

war years, average American government support for scientific research increased more than tenfold, from $48 million to about $500 million, going from at most 18 percent of total national research expenditure to approximately 83 percent (28). Equally significant was the strong concentration of research in a comparatively small number of university and industrial laboratories:

Some 66% of wartime contract dollars for R&D went to only 68 corporations, some 40% to only ten. OSRD spent about 90% of its funds for principal academic contractors at only 8 institutions, about 35% at the MIT Radiation Laboratory alone. And more than nine out of ten of the contracts from OSRD and the military granted the ownership of patents deriving from this funded research to the contractors, not to the public (29).

At the end of 1945 OSRD reported it had given wartime contracts totalling $117 million to MIT, another $83 million to Cal Tech, $31 million to Harvard, and $28 million to Columbia (30). By way of comparison, a survey of research expenditure at the thirteen leading American academic physics institutions showed that for the 1939—40 academic year the highest total annual research expenditure for an individual physics department was only $39,000 (31). After the war, "The Office of Naval Research [ONR] alone [had] university contracts in physics . . . at a current rate over $10 million a year" (32) and had a total grant budget for all sciences of approximately $20 million by 1949 (33). The sources of ONR's funding are particularly interesting, as it is by no means clear how much of it actually came from the Navy's own coffers:

The Atomic Energy Commission [from its foundation in 1946] then used ONR as its preferred mechanism for the support of university-based research [as it was barred by its act of enactment from making such grants itself — PKH] . . . Rear Admiral Luis de Florez [of ONR] . . . was said to have persuaded the Bureau of Aeronautics, his old agency, to transfer in $32 million. As it was revealed that he was then already or soon thereafter to become the director of technological research at the CIA, it is possible now to doubt the generosity of the Bureau of Aeronautics (34).

As the Office of Naval Research was from 1946 the leading contractor for fundamental university research, the above statement, if correct, would indicate that a decision may have been made at a very high level to try to keep the university research base of the country intact and in close liaison with the military.

Such a conclusion is also suggested by President Truman's reported opposition to the termination of the wartime Radiation Laboratory of the radar project, and his behind-the-scenes role in its subsequent reincarnation as MIT's Research Laboratory of Electronics (RLE). According to the Institutes's physics chairman, John C. Slater:

> Truman felt that by such a termination a great deal of scientific knowledge would be lost . . . . He hoped that in the course of time the armed services would be able to set up contracts [as ONR would soon be doing] to keep the sort of work which the OSRD had carried on going in peacetime. In order not to have all scientific work lapse while these contracts were being set up, he gave directions to the OSRD to have certain projects [including the Rad Lab] continue until June 30, 1946 . . . . It became at once obvious that this directive on his part, and our plans for the RLE, complemented each other perfectly.

Pending the actual signing of contracts, MIT agreed to set up a revolving fund so that the physics and engineering staff that formed RLE could take "advantage of a situation that existed in the Radiation Laboratory as a result of the war . . . [in which talented] young men from all over the country had been drawn in" (35). The existing staff, and others attracted from Los Alamos and other wartime laboratories, could now be offered lucrative research associateships to maintain the concentration of talent at MIT. Of the professors who set up RLE, two — Julius Stratton and Jerome Wiesner — subsequently became presidents of MIT, while a third, Albert G. Hill became vice-president in charge of research at MIT and later Chairman of the board of the controversial Draper military instrumentation laboratory, which in the early seventies was made nominally independent of the Institute (36). Another indirect result of the presence of the Rad Lab was the setting up at the end of the 1940s of the Lincoln Laboratory, with Hill as director, to deal with the more applied military research previously handled by RLE.

Along with MIT, Cal Tech, Chicago, and Berkeley, another major university beneficiary of wartime research was Harvard — which had contained the Radio Research Laboratory (RRL), in the process acquiring considerable money, contacts, equipment, and expertise, especially in microwave electronics. The latter eventually led to R. V. Pound and E. M. Purcell's work on nuclear magnetic induction, as well as the upgrading of what became the Harvard School of Engineering and

Applied Physics (headed by former RRL theory-group director John H. Van Vleck). Like MIT's RLE, after the war Harvard's Cruft Laboratory was almost wholely militarily financed, as were also the Columbia Radiation Laboratory, Duke University's Air Force-funded Microwave Laboratory, Brooklyn Polytechnic's Microwave Research Institute, and Stanford's Microwave Research Laboratories (37).

The concentration of wartime and postwar university research led to an enhancement of the most favored institutions in the sciences at the expense of others less favored — including, at first, such top-rung institutions as Princeton, Cornell, and Johns Hopkins (though the latter eventually obtained the wartime contract for the proximity fuse, out of which effort came its Applied Physics Laboratory). It was perhaps of symbolic significance that just after the war DuBridge, who had headed the MIT Rad Lab during the war, became president of Cal Tech, second only to his previous institution among the leading university war contractors. As Lord Bowden pointed out in a Royal Society of London discussion document:

> The scale on which [major] American universities developed [in this period] is still not understood or comprehended in the United Kingdom. For many years MIT alone disposed of greater resources for its research programme than all the English universities put together . . . (38).

Another feature of the postwar world was the continuation in one form or another of most of the major research laboratories of the Manhattan Project, often within — or in close cooperation with — particular universities. In the Chicago area, the wartime Metallurgical Laboratory of the Manhattan Project was succeeded by the permanent establishment of the Argonne National Laboratory. The Oak Ridge Laboratory was likewise maintained in Tennessee; and the Lawrence Radiation Laboratory continued at the University of California at Berkeley; as did the Los Alamos Laboratory, which had been nominally associated with the university. The official history of the Lawrence Laboratory notes that after the war, in 1945 and 1946:

> The Manhattan Engineering District [of the Department of War] generously supported the Rad Lab's conversion to peacetime research . . . and authorized the completion of

the 184-inch synchrocyclotron [costing the MED $170,000] and the construction of an electron synchroton [costing $230,000 plus $203,000 for surplus equipment from Oak Ridge]. . . . [Professor Luis] Alvarez got support for preliminary work on a linear accelerator [estimated to cost $5.5 million]. . . . The semi-annual budget [for the lab] from the MED amounted to $1,370,000 (39).

At this point complications arose, as the new Atomic Energy Commission (AEC), which took over from the MED on January 1, 1947, was prohibited by its act of enactment from giving outside grants for research. Moreover, James B. Fisk, its director of research (later president of Bell Laboratories, and then the member of the MIT Corporation "most closely associated with picking out the last several presidents of MIT" [40]), was known to be hostile to spending AEC money on university accelerators. He was eventually persuaded, and in October 1947 the AEC allocated $15 million for such equipment. "To complete the loop, Kenneth Pitzer, professor of chemistry at Berkeley succeeded Fisk. By the end of 1948 AEC research policy had been shaped to assure the future of fundamental nuclear science [especially at Berkeley]" (41).

A major industrial beneficiary of wartime research largess was the American Telephone and Telegraph Company (AT&T), parent company of both Western Electric and Bell Laboratories, which in 1939 had government contracts amounting to only $200,000 (about 1% of its total research expenditure); by 1944 this had risen to $56 million (about 81% of its total research expenditure) (42). Early in the war there had been considerable disagreement between MIT and Bell administrative personnel over which institution was better qualified to host the Rad Lab, and indeed all through the war there was further dispute about the slowness with which Bell handed over information that might be of future commercial value. On the other hand, without the wartime expertise in the properties of semiconductor crystals that Bell gained through other Rad Lab contractors like Purdue and the University of Pennsylvania, it might have been much less likely to develop the transistor. Along with AT&T and its subsidiaries, the list of top OSRD corporate war contractors includes the industrial laboratories of Du Pont, RCA, Eastman Kodak, and General Electric. This narrow concentration of research funding has since been justified in retrospect by "policies of buying the best and causing as little disruption to the status quo as possible" (43) — but it and similar postwar arrangements were also to lead to the radical reorientation of university and industrial resources later identified by President Eisenhower with the rise of a military-industrial complex.

In Britain, Malvern — the wartime site of the Telecommunications Research Establishment and the so-called British Branch of the Radiation Laboratory — was to reemerge as the Royal Radar Establishment (later RSRE), undoubtedly the largest electronics laboratory in the country. British nuclear research was to be heavily concentrated after the war in the immense Atomic Energy Research Establishment at Harwell, with atomic weapons research later farmed out to Aldermaston. These laboratories were run on a scale that completely dwarfed prewar support for the National Physical Laboratory (44). They were also far larger than comparable university laboratories, usually by at least a whole order of magnitude, and did not normally channel substantial subcontracts through the universities. In this sense, the British military-industrial complex has managed to maintain itself for the most part outside the universities and, on the whole, to keep a lower profile.

3. Debates about the Terms of the Alliance

Despite ex post facto reminiscences depicting the wartime alliance of scientists and the military as sweetness and light, it is clear that there were often tensions of varying severity between the two groups, ultimately about who would set the terms of their common understandings. In the Manhattan Project, there was a good deal of resentment about General Groves's policies of severe secrecy, compartmentation of information, and security screenings. More than that, there was tension and insecurity about the relative positions and prerogatives of the various participants. Some of this shows through starkly in a memorandum written in August 1943 by a military intelligence agent, Captain Pier de Silva, suggesting how the War Department should deal with a compromising incident involving Oppenheimer:

It is the opinion of this officer that Oppenheimer is deeply concerned with gaining a worldwide reputation as a scientist, and a place in history, as a result of the [Manhattan] project. It is also believed that the Army is in the position of being able to allow him to do so or destroy his name, reputation, and career, if it should choose to do so. Such a possibility, strongly presented to him, would possibly give him a different view of his position with respect to the Army, which has been heretofore, one in which he has been dominant because of supposed essentiality. If his attitude should be changed by such an action, a more wholesome and loyal attitude might, in turn, be injected into the lower echelons of [scientist] employees (45).

There is no record that this was actually and explicitly done. However, it is clear that the security screenings and postwar loyalty investigations accomplished – implicitly – the same general results. They tended to give the government, the military, and the intelligence services the whip hand in their dealings with scientists.

With the postwar spread of militarily sponsored research both inside and outside the universities, questions about one's prior political associations or reliability could at the very least endanger one's sources of funding or participation on government advisory committees, if not indeed one's actual employment. From 1947 there was also President Truman's sweeping security order requiring a loyalty screening for all employees of all federal agencies. In this period there was sharp controversy about the sort of political criteria applicable to senior scientific civil servants – especially to Edward Condon, director of the National Bureau of Standards. Similar issues arose over Senator McCarthy's investigation in 1954 into the U.S. Army Signal Corps' main in-house electronics laboratory at Fort Monmouth, New Jersey.

The ensuing public controversies helped shape the relative positions, roles, and prerogatives of the participating elites – especially of the senior scientific elite of people like Condon and Oppenheimer, who were in some ways its most creative and indispensable component, but who entered it on uncertain terms that were defined only in the political controversies that followed. Although the scientific elite at various points sought the same sort of postwar role in defining weapons policy that the Office of Naval Research by then possessed in relation to university research priorities, they were at every point easily blocked by the military (46). This indicates their relative weakness and lack of control over resources, as compared to their military and industrial counterparts. Like other elites lacking direct control over economic resources and institutionalized political positions, their power rested ultimately on the potential for political mobilization of their own constituency. But this was not something they would easily resort to, for it would have had the somewhat contradictory effect of exposing their private dealings to rank-and-file questions, and hence eroding their elite positions. This led to the somewhat paradoxical situation in which, after the withdrawal of Oppenheimer's security clearance, scientific leaders publicly bewailed the dire effects of a possible scientists' strike or a de facto break-up of the alliance — while privately doing everything they could to see that this would never happen (47).

The development of such a successful and long-lasting postwar alliance seems, in some respects, especially surprising in view of the long-term opposition by American scientists to what was previously seen as government interference in science, which had led only a few years earlier to the demise of the Science Advisory Board. As late as 1943, when Senator Harley Kilgore put forward a bill to set up a government Office of Technological Mobilization (later to transmute into a proposal for a National Science Foundation) he was opposed by Frank Jewett of Bell Labs and the National Academy, on the grounds that this

> would completely revolutionize our entire American age-old concept and, by placing vast powers in the hands of a federal bureaucracy, would set the stage for complete domination of the life of the nation by a small group of federal officers and bureaucrats.

Jewett added that he had received numerous letters from fellow scientists that "reflected an unalterable opposition to being made the intellectual slaves of the state . . . ." Nevertheless, it is arguable that the American scientific elite in their various struggles — for the prolongation of a government-sanctioned NRC after the First World War, for a National Science Endowment in the mid-twenties, for the Science Advisory Board in the mid-thirties, and eventually for a certain form of National Science Foundation in the forties — were by no means opposed to federal sponsorship of science (48).

The debate about the structure of the proposed National Science Foundation (NSF) in the 1940s polarized around the July 1945 proposals of Kilgore and Bush, respectively. The key differences were that Bush's proposal would create a foundation virtually independent of presidential control, would make no attempt to distribute funds on anything like a geographical basis, and would allow grantees to retain patent right . . . [and in general] sought to limit the role of the government to merely that of paying the bill (49).

Both sides saw the issues through the lenses of their own ideologies. To Kilgore and other political figures — including President Truman, who vetoed one version of the Bush proposal two years later — the dispersal of government funds could only be done by the elected representatives of the people (or those they delegated for this task), and any patents arising out of government-financed research should rightly belong to all the people. On the other hand, Bush and the scientific elite saw Kilgore and other congressional spokesmen of his point of view as "critics [who] sought to make fundamental changes in the structure of American science" (50) — by exposing it to the whims of "politicians," and to acrimonious public debates about its place within overall social (not just scientific) priorities. This was not entirely wrong-headed, for if science became politicized — as it did in fact become in the postwar debates, particularly over atomic energy and weapons policy — the results could be extremely destructive, as they no doubt were in the Condon and Oppenheimer cases. A science buffeted by such political winds could be severely damaged.

The scientific elite therefore sought a National Science Foundation that would operate mostly outside the glare of public debate, insulated by a board of leading scientists, dispersing funds according to scientific — not social — priorities. Although entirely sensible within its own terms, this proposal was seen as somewhat self-serving: in effect, the scientific elite would be distributing public funds, for the most part to itself and its own institutions, and with only the most minimal public controls. The crucial issues were, of course, the purpose of government support and the related issue of its control. If the purpose was primarily to aid the main centers of academic science along the lines of their own programs, then the scientific leadership could exercise control. But if the intention was to stimulate science to meet the all-around needs of society — as the social-function-of-science movement in Britain had advocated — then resources would have to be dispersed far more widely, and decision-making would have to involve wider constituencies, with ultimate control in the hands of elected representatives of society as a whole. To elite spokesmen like Bush the first alternative was acceptable, the second a kind of slavery.

During five further years of wrangling, atomic energy and weapons were taken over by the AEC (created in 1946); medical research was put under the National Institutes of Health; and conventional weapons research, along with anything that might be remotely relevant to it, was claimed by ONR and the organs of the reorganized Department of Defense. For these five crucial postwar years, the main finance for even "basic" university research was provided by ONR, followed by other military outlets (after 1950) and the partially militarized AEC. Once the situation had congealed in this way — reflecting an imbalance of power between the military and the scientists — it was not significantly changed even by the creation of the NSF, which had far less to spend, and was able to claim only a distinctly secondary position in the financing of scientific research.

"An assumption that had some currency in the early years was that the NSF would take over in the form of transfers the basic research already being performed by the AEC, the Department of Defense, and various other agencies of the government" (51). This did not happen, because neither the military, nor the scientific elite who were doing well out of military contracts, particularly wanted it to happen (52). Instead, in Executive Order number 10521, President Eisenhower in March 1954 gave to federal agencies the power to "conduct and support . . . basic research in areas which are *closely related* to their missions" (53). The NSF, after its formation in 1950, could of course have pressed more strongly to include all university basic research under its remit, but its first director, Alan T. Waterman — a former Yale physicist, who had until recently been chief scientist for ONR — thought it unwise to do so.

In view of the previously mentioned hue and cry about slavery to politicians, it may seem surprising that military patronage was accepted relatively calmly. One reason for this was that the military elite, like the scientific elite, were well used to making their arrangements in private, away from the tensions of public scrutiny. Despite some disagreements, the two had worked reasonably well together in the wartime OSRD and knew they could do business. In fact, there were a number of occasions both during and after the war, on which the elite of scientific administrators took the side of the military, even against what may have been the bulk of their own ostensible constituency. One such occasion was in 1945 when the scientific Panel (including Oppenheimer, Enrico Fermi, E. O. Lawrence, and A. H. Compton) of the Manhattan Project's Interim Committee — appointed to decide, amongst other things, if the atom bomb should be used without warning against Japan — overruled the negative sentiments of rank-and-file atomic scientists as expressed in the Franck Report. Another occurred the following year when Bush, Conant, Lawrence, and Oppenheimer spoke out in favor of military control of atomic energy, as expressed in the May-Johnson Bill then before Congress, against the preponderant opinion within the scientific community. (This was admittedly during a period when the disposition of the equipment and assets from the wartime scientific laboratories was still being decided, not to mention the continuation of military and MED support for particular laboratories and projects.) Against the grain of rank-and-file scientific opinion, Lawrence also supported at least tacitly the decision to remove Oppenheimer's security clearance, and A. H. Compton two years later wrote a justification of that decision. When, during the first few months of the Los Alamos Laboratory's existence, Edward U. Condon challenged the military's policies of compartmentation of information, the lab's director, Oppenheimer, did not support him, and Condon resigned shortly after. Indeed, "unlike most scientists — who distrusted or disliked [General] Groves — Bush, Conant, A. H. Compton and Oppenheimer had an excellent relationship with him and were generally sympathetic to his point of view" (54). This seems to illustrate a more general phenomenon in which diverse elites doing business together over a prolonged period acquire a kind of corporate character (or common interest in maintaining privacy and power) that in some ways sets them off from — even, occasionally, in opposition to — the constituencies that they ostensibly claim to represent. In this case, "The wartime scientific leadership prided itself on having made possible a partnership with the military that was greatly responsible for the success of the Manhattan Project. It hoped that the foundations of this partnership would be cemented after the war" (55).

At first, the latter objective seemed easily realized, particularly as the group of men who formed ONR were especially respectful of scientists and the benefits that could accrue from even their basic research. These young naval officers seem consciously to have seen themselves as being uniquely placed to facilitate the scientists' work. The "Bird Dogs" had, nevertheless, to work fairly hard in the year after the end of hostilities to convince university administrators to accept research grants from a military source in peacetime (56). As late as 1948, resistance to military funding of universities came "chiefly from university administrators seriously concerned for their institution's independence, while their physicists, first and foremost among the scientists, pressed for such facilities as only the military could or would provide" (57). ONR was also aided in its dealings with academic administrators and with rank-and-file scientists in no small way by the willingness of the scientific elite in almost every field to serve on grant advisory committees.

For the most part, the scientists who had dominated the university, industrial, and foundation elites of the 1930s were men of white, Anglo-Saxon, and Protestant professional backgrounds. These men were joined in the war, especially in the Manhattan Project, by a number of others — such as Oppenheimer, Fermi, Edward Teller, I. I. Rabi, Hans Bethe, and John von Neumann — of Jewish and/or émigré background. Considering the tensions about the entry of Jews and émigrés into the staffs of American universities in the thirties (58), it is remarkable how well the two groups were able to work together in the forties. There was, nevertheless, a strong tendency (even more pronounced in Britain than in America) to appoint to the leadership of OSRD and the radar project, at first, almost exclusively men of mainstream background. This was justified at the time by the previous experience of this group in university physics administration (then more or less closed to the others) and by military and Civil Service regulations forbidding the employment of aliens. It was only somewhat later that the "ethnics" began filling more visible roles within the — at first more peripheral and speculative — atomic weapons program. Because of their comparatively high-level positions within that program and perhaps in part because they were less able to return to similarly high-level administrative posts in universities in 1945 than were their more mainstream counterparts, what I'll call the composite Jewish/émigré group continued to maintain a high profile in the postwar atomic programs. They were therefore destined to play a prominent role — for the most part, as defendants — in the subsequent security investigations. Thus, in the Oppenheimer case the principal protagonist was a native-born German-Jew, defended most strongly by three others of similar ethnicity: the Austro-Hungarian-born Rabi, the German-born Bethe, and the former chairman of the AEC, David Lilienthal. (In this case, one of his principal accusers was a Hungarian-born Jew, Edward Teller, but the other scientists who testified on Teller's side were not of Jewish background.) The continuing security investigations of leading Jewish scientists, as well as the Fuchs case in Britain and the case of the "atom spies" — Julius and Ethel Rosenberg — in America, fanned fears within the Jewish community of the same sort of anti-Semitic purge that had marked the last years of Stalin in the Soviet Union. Among the mainstream of the scientific community, there were similar fears of hysteria against scientists (59).

For many of the émigrés, including Leo Szilard and Fermi, their participation in the atomic project was their route into the American scientific center. For von Neumann and Teller it was a route into the administrative elite. Even for the mainstream of the prewar scientific elite, there was no disposition to go back to the comparatively penurious conditions of the 1930s. One should not underestimate the tensions of the depression as a force propelling them to accept continued government and military largess. Indeed, considering that the American Institute of Physics was formed in part to counter arguments against the social usefulness of science (60), having proved themselves in the war the physicists were all too ready to accept the rewards that they felt they had earned.

There was, nevertheless, some significant opposition, both to channeling public support for science through the military and, much more widely, to the various restrictions that seemed to go with this. Even at that time there were some who suspected that he who pays the piper eventually calls the tune, so that even the relatively beneficent support of ONR might be transformed by slow stages into military control. The fullest debate about such issues was to emerge during the Vietnam War, though as early as 1947 Albert Einstein publicly questioned whether "funds raised for these purposes from the taxpayer should be entrusted to the military," and asserted:

> Every prudent person will certainly answer: "No!" For it is evident [within the context of the prevailing mandarin ideology] that the difficult task of the most beneficent distribution should be placed in the hands of people whose training and life's work give proof that they know something about science and scholarship (61).

Equally serious objections were raised at the time by Cornell physicist Philip Morrison, who worried that "science will have been bought by war, on the installment plan"; by MIT mathematician Norbert Wiener, who did not want his work to be employed for military ends; and by Harvard's president James Conant, who was concerned about the entry of secret research onto the campuses in peacetime. In the gathering atmosphere of cold war and security-mindedness, all three would subsequently face innuendos about their loyalty (62). Privately, MIT's president Karl Compton confided that his Corporation was concerned that government and military research would prove a most unreliable source of finances. And there was more general worry in the universities about the phrasing of contracts and the pressures, usually implicit, for degrees of matching university support (63). There was also the campaign of rank-and-file atomic scientists against military control of atomic energy, and against restrictions arising from science's continuing mobilization.

As late as 1949, DuBridge still worried over the danger of "regimentation of [American] science, its subservience to immediate military ends, or undue secrecy restrictions," and he insisted:

> There will be uneasiness until it is more clearly evident that the people of this country through *non-military* agencies of their government and also as individuals and through corporations, foundations and other agencies will insist on and adequately support a strong and free science [italics his].

He added that

> a major aspect of postwar physics [has been that] there are still restrictions being imposed upon the freedom of science and of scientists which are neither necessary nor

desirable — which are indeed inimical not only to the welfare of science but also to the welfare of the nation. The first of these restrictions is the set of secrecy rules — mistakenly referred to as security rules — with which science was necessarily enveloped during the war but which have not yet [in 1949] been adequately lifted (64).

Five years later, Lloyd Berkner, president of Associated Universities (the consortium of universities that initiated the Brookhaven National Laboratory) and a strong supporter of the university-military alliance, vividly denounced what he called "the mania for secrecy" and insisted that "the protection of our whole democratic system, of which academic freedom is but a part, calls for a searching inquiry into this creeping infection and measures for its prompt control" (65). Unfortunately, the vigor of debate within the American scientific community over the loyalty and security investigations — that is, the *terms* of their alliance with the military — was, by a process of cognitive dissonance resolution familiar to psychologists, to obscure from them the somewhat larger question of the propriety of being in such an alliance in peacetime at all (66).

This is not, of course, to say that the security restrictions and investigations were not real and did not have enduring individual (and social) repercussions. Condon was forced to resign his directorship of the National Bureau of Standards under pressure, and for the next decade was more or less black-listed from both government and university service. Similar tragedies befell a substantial number of other scientists, including Condon's friend and former collaborator at Princeton, Ronald Gurney, who was denied a security clearance at Brookhaven, hounded out of various positions elsewhere, and essentially destroyed by the experience. He died of a stroke in New york in 1953, virtually destitute (67). On the eve of Senator McCarthy's investigation of the U. S. Army Signal Corps Laboratory at Fort Monmouth, the Army suspended forty-two of its scientists and engineers, mostly without charges or hearings (68). Oppenheimer's hearing before the AEC was in some ways the nadir of this whole process — for if the government's most eminent scientific advisor could be defamed and driven from government service, then scientists were little more than dispensable hired hands. True, the Army and the scientists (at least those who were not suspended) did, to a certain extent, make common cause against — and actually defeat — McCarthy, but the results of the Oppenheimer decision were not reversed and a smoldering resentment continued throughout the fifties in much of the scientific community. This was somewhat alleviated only in December 1963 with the presentation of the Fermi award to Oppenheimer by President Johnson, though the scientist's security clearance was not in fact restored (69).

To be sure, even the elites of science and the military have in general significantly different aims and values, which overlap only at certain times and only in limited respects. It is doubtful whether, even as late as 1940, they would have been able to work so closely together in any but such a potentially all-encompassing national emergency. It is also doubtful whether their cooperation in the postwar era could have continued in the way that it did — and as long as it has — without the continuing atmosphere of cold war and mobilization. Some have argued, on the other hand, that rather than the cold war providing the cement and rationalization for the continuance of this allliance, it is instead the forces behind the alliance that find it in their interest to maintain the cold war (70).

In directing attention to the elites that formed and mediated such an alliance, I do not wish to ignore the social forces behind it. Such a development could hardly have taken place had it not in some ways been in the interests of the military, of key sectors of industry, and of at least certain forces within (and on the boards of) the universities most centrally involved. In the military emergency the alliance obviously served the immediate interest of the nation as a whole, though whether it continues to do so in the longer term is much more problematic. Clearly, without cold war appropriations for the military, military research would be decimated, and could no longer serve as a major funder for university research. Whether this would mean less — or more — funding for university research is of course another matter, as the possibility could thereby emerge of converting what were military appropriations to civilian purposes. In the latter 1940s, with the National Science Foundation proposal stalled, the scientific elite did not place much credibility in such an outcome and for the most part readily adapted itself to continued military funding — especially as the pill was sweetened by the establishment of advisory panels on which the scientists seemed to exercise substantial influence. For them, the proof of this pudding was clearly in the eating: their ability to consume substantial grants for their own institutions. Even so, as Harvey Sapolsky has perceptively observed, "[by] their enthusiasm for research subsidies and their willingness to accept whatever rationales appeared to be effective in gaining [them, the scientific elite] have produced precisely the impact upon science they sought to avoid – its permanent mobilization" (71).

Acknowledgment

This work was begun as part of the International Project on the History and impact of Solid State Physics, supported in the U.K. by the Leverhulme Foundation and the British Academy, with organizational support through the American Institute of Physics Center for History of Physics.

Notes

1. "The whole thing is repugnant to the ordinary civilian-life ways of scientists – the restrictions under which they have to operate and the cellular structure. The uniform experience in talking to all men who have given a lot of time and effort to OSRD is that they want to get out of this thing . . ." (testimony of Dr. Frank Jewett, Vice-President, Bell Telephone Laboratories, in U.S. Congress, House Select Committee on Postwar Military Policy, *Hearings, Research and Development* [Washington, D.C.: GPO, 1945] p. 58).
2. On this point see Paul K. Hoch, "Toward a Sociology of Discovery: Migration and the Generation of New Scientific Ideas," *Minerva* **25** (Autumn 1987), 209–237.
3. See especially Yaron Ezrahi, "The Authority of Science in Politics," in Everett Mendelsohn and Arnold Thackray (eds.), *Science and Values* (New York: Humanities Press, 1974), pp. 215–251.
4. Alice K. Smith, *A Peril and a Hope: The Scientists' Movement in America, 1945–47* (Chicago: University of Chicago Press, 1965); Morton Grodzins and Eugene Rabinowitch (eds.), *The Atomic Age: Scientists in National and World Affairs* (New York: Basic, 1963); Robert Gilpin, *American Scientists and Nuclear Weapons Policy* (Princeton: Princeton University Press, 1962); William R. Nelson, *Case Study of a Pressure Group: The Atomic Scientists* (Ann Arbor: University Microfilms, 1966).
5. Norbert Elias, Herminio Martins, and Richard Whitley (eds.), *Scientific Establishments and Hierarchies* (Dordrecht: Reidel, 1982).

6. Burton J. Bledstein, *The Culture of Professionalism* (New York: Norton, 1976); Everett Mendelsohn, "The Emergence of Science as a Profession in Nineteenth-Century Europe," in Karl Hill (ed.), *The Management of Scientists* (Boston: Beacon, 1964), pp. 3—48; M. S. Larson, *The Rise of Professionalism* (Berkeley: University of California Press, 1977); John J. Beer and W. David Lewis, "Aspects of the Professionalization of Science," in Kenneth S. Lynn et al. (eds.), *The Professions in America* (Boston: Beacon, 1963), pp. 110—130; Daniel J. Kevles, "The Study of Physics in America, 1865—1916" (Ph.D. dissertation, Princeton University, 1964).
7. Ferdinand Lundberg, *The Rich and the Super-Rich* (New York: Lyle, Stuart, 1968), pp. 146 ff; Max Dorian, *The Du Ponts: From Gunpowder to Nylon* (Boston: Little, Brown, 1961).
8. John J. Beer, "Coal Tar Dye Manufacture and the Origins of the Modern Industrial Research Laboratory," *Isis* **49** (1958), 124 ff; Mortimer D. Fagen, *A History of Engineering and Science in the Bell System: The Early Years, 1875—1925* (New York: Bell Laboratories, 1975); Kendall Birr, *Pioneering in Industrial Research: The Story of the General Electric Research Laboratory* (Washington, D.C.: Public Affairs Press, 1957); Spencer R. Weart, "The Rise of 'Prostituted' Physics," *Nature* **262** (1976), 13—17; Paul K. Hoch, "The Rise of Physics Laboratories in the Electrical Industry," *Physics in Technology* **16** (1985), 177—183.
9. Ronald C. Tobey, *The American Ideology of National Science, 1919—1930* (Pittsburgh: University of Pittsburgh Press, 1971), p. 37.
10. National Academy of Sciences, *Annual Report for 1918* (Washington, D.C.: NAS, 1918), pp. 40—41.
11. In a private letter to a friend, Hale declared the world war emergency as "the greatest chance we ever had to advance research in America" (quoted by Helen Wright, *Explorer of the Universe* [New York: Dutton, 1966], p. 288).
12. The diary is in the G. E. Hale Papers, Archives of the California Institute of Technology.
13. Tobey, *op. cit.*, 1971 (9), p. 41. See also Robert M. Yerkes (ed.), *The New World of Science: Its Development During the War* (New York: Century, 1920).
14. For example, see Fritz Ringer, *The Decline of the German Mandarins* (Cambridge, Mass.: Harvard University Press, 1969); Paul Forman, "Scientific Internationalism and the Weimar Physicists: The Ideology and its Manipulation in Germany after World War I," *Isis* **64** (1973), 151—180.
15. Tobey, *op. cit.*, 1971 (9), p. 47.
16. *Ibid.*, pp. 50—53.
17. A. Hunter Dupree, *Science in the Federal Government: A History of Policies and Activities to 1940* (Cambridge, Mass.: Harvard/Belknap Press, 1957), p. 358; see also Karl T. Compton, "Program for Social Progress," *Scientific Monthly* **44** (1937), 6 ff.
18. Alex Roland, *Model Research: The National Advisory Committee for Aeronautics, 1915—1918*, 2 vols. (Washington, D.C.: GPO, 1985).
19. J. L. Penick, Jr., C. W. Pursell, M. B. Sherwood, and D. C. Swain (eds.), *The Politics of American Science 1939 to the Present* (Cambridge, Mass.: MIT Press, 1965), p. 9.
20. Elizabeth Hodes and Robert Kargon, "Karl Compton, Isaiah Bowman, and the Politics of Science in the Great Depression," *Isis* **76** (1985), 305—306.

21. Details are available in the Henry Barton Papers, American Institute of Physics (hereafter, AIP); and see Henry Barton, "The Story of the American Institute of Physics," *Physics Today* **9** (January 1956), 56—66; Karl Compton, "The Foundation of the AIP," *Physics Today* **5** (February 1952), 4—7. See also AIP, *Physics in Industry* (New York: AIP, 1937).
22. Dupree, *op cit.*, 1957 (17), p. 370.
23. Quoted by Dupree, *ibid.*, p. 371.
24. Daniel J. Kevles, *The Physicists: The History of a Scientific Community in Modern America* (New York: Vintage, 1979), pp. 308 ff.
25. Lee A. DuBridge, "The Effects of World War II on the Science of Physics," *American Journal of Physics* (1949), 274.
26. Gary Werskey, "Science and War in the Twentieth Century," in *War in Our Own Day* (Milton Keynes: Open University Press, 1973), pp. 32ff estimates that U.S. and U.K. military and militarily related ("prestige") R&D is of the order of 50 percent of total public and private R&D, a percentage that is much higher than in West Germany or Japan. This is in one respect the material basis for a military-industrial-scientific complex.
27. Kevles, *op. cit.*, 1979 (24), p. 370; *Institute of Radio Engineers Proceedings* **50** (1962), 540.
28. President's Scientific Research Board, *Science and Public Policy: A Report to the President, 1947*, 5 vols. (Washington, D.C.: GPO,1947), vol. 1, 10.
29. Kevles, *op. cit.*, 1979 (24), p. 342; Kevles cites Carroll W. Pursell, Jr. (ed.), *The Military-Industrial Complex* (New York: Harper and Row, 1972), pp. 165—169.
30. James P. Baxter III, *Scientists against Time* (Boston: Little, Brown, 1946), p. 456.
31. V. Bush, *Science, The Endless Frontier*, 2nd ed. (Washington, D.C.: Public Affairs Press, 1960), pp. 128—129; cited by Daniel S. Greenberg, *The Politics of American Science* (Harmondsworth, Middlesex: Penguin, 1969), p. 136.
32. DuBridge, *op. cit.*, 1949 (25), p. 278.
33. Penick et al., *op. cit.*, 1965 (19), pp. 24, 185; U.S. Navy, *Annual Report of the ONR* (1949), p. 1.
34. Harvey M. Sapolsky, "Academic Science and the Military: The Years since the Second World War," in Nathan Reingold (ed.), *The Sciences in the American Context: New Perspectives* (Washington, D.C.: Smithsonian Institution Press, 1979), pp. 384—385; Sapolsky cites the comments of Adm. de Florez in "Transcript of the Ninth Meeting of the Naval Research Advisory Committee" (Washington D.C., June 1949), p. 75. See also reports of de Florez's CIA position in the *New York Times*, January 11, 1976.
35. John C. Slater, *Solid Sate and Molecular Theory: A Scientific Biography* (New York: Wiley/Interscience, 1975), pp. 223, 220.
36. Interviews with Albert G. Hill and Julius Stratton, August 1982. On the Draper Lab controversy see Dorothy Nelkin, *The University and Military Research* (Ithaca, N.Y.: Cornell University Press, 1972).
37. Paul Forman, "Behind Quantum Electronics: National Security as a Basis for Physical Research in the United States, 1940—1960," *Historical Studies in the Physical and Biological Sciences* **18** (1987), pp. 149—229.
38. Lord Bowden, "Effects of World War II on Education in Science," *Proceedings of the Royal Society of London*, ser. A, **342** (1975), 501.
39. J. L. Heilbron et al., *Lawrence and His Laboratory: Nuclear Science at Berkeley*

1931—1961 (Berkeley: Office of History of Science and Technology, University of California, 1981), pp. 47—49.

40. Slater, *op. cit.*, 1975 (35), p. 170.
41. Heilbron et al., *op cit.*, 1981 (39), p. 49.
42. U.S. Congress, Senate Subcommittee on War Mobilization, *The Government's Wartime Research & Development, 1940—44; Part II — Findings and Recommendations*, 79th Congress, 1st Session, 1945, pp. 21—22; excerpted in Penick et al., *op cit.*, 1965 (19), pp. 99—100.
43. Details about the disputes between MIT, Bell, and the Rad Lab can be found, e.g., in the Radiation Laboratory Papers at the Federal Archive and Record Center, Waltham, Mass.; the Vannevar Bush Papers in the Library of Congress; and the Karl Lark-Horowitz Papers at the American Institute of Physics. For corporate war contractors see Baxter, *op cit.*, 1946 (30), pp. 456—457; and Carroll Pursell, "Science Agencies in World War II: The OSRD and Its Challengers," in Reingold, *op. cit.*, 1979 (34), p. 364, from which the quote in the text has been drawn.
44. R. A. Smith, "Physics at the Radar Research Establishment, Malvern," *Proceedings of the Royal Society*, ser. A. **235** (1956), 1. Between 1931 and 1961 the number of scientists and engineers in the U.K. scientific Civil Service multiplied nearly 15 times over, and both AERE-Harwell and the Royal Aircraft Establishment had "a greater scientific strength than the whole of the pre-war scientific Civil Service" (R. V. Jones, "Research Establishments," *Proceedings of the Royal Society* ser. A. **342** [1975], 486). See also Sir John Cockcroft (ed.), *The Organization of Research Establishments* (Cambridge: Cambridge University Press, 1963).
45. Atomic Energy Commission, *In the Matter of J. Robert Oppenheimer. Transcript of Hearing before Security Board* (Washington, D.C.: GPO, 1954), p. 275.
46. Sapolsky, *op. cit.*, 1979 (34), p. 391.
47. On this point see especially Joseph Haberer, *Politics and the Community of Science* (New York: Van Nostrand, 1969), chaps. 10—12; testimony of Vannevar Bush, *U.S. Congress, House Committee on Government Operations, Subcommittee on Military Relations, Hearings, Organization and Administration of the Military Research and Devellopment Programs* 83rd Congress, 2nd Session, 1954, pp. 843 ff; V. Bush, "If We Alienate Our Scientists," *New York Times Magazine* June 13, 1954, pp. 9, 68—71.
48. Frank Jewett to V. Bush, May 27, 1943, Bush file, Jewett folder, Carnegie Institution of Washington Archives; Jewett to Bush, September 2, 1942, U.S. National Archives, Record Group 227, Central Classified file, Kilgore Committee, July—December 1943. These debates are well documented in, e.g., Kevles, *op. cit.*, 1979 (24), pp. 110—118, 139—140, 185—187, 252—262, 344—366; Tobey, *op cit.*, 1971 (9), esp. pp. 34—61; Penick et al., *op. cit.*, 1965 (19), esp. pp 102—119; Greenberg, *op. cit.*, 1969 (31), esp. pp. 95—98, 146 ff; and D. J. Kevles, "The NSF and the Debate over Postwar Research Policy, 1942—45," *Isis* **68** (1977), 5—26.
49. Carroll Pursell, "Science and the Government Agencies," in D. D. Van Tassel and Michael G. Hall (eds.), *Science and Society in the United States* (Homewood, Ill.: Dorsey, 1966), p. 243; see also V. Bush, *Science, The Endless Frontier* (Washington, D.C.: Public Affairs Press, 1945), p. 50.
50. Pursell, *op. cit.*, 1966 (49), p. 242.

51. NAS, *Federal Support of Basic Research in Institutions of Higher Learning* (Washington, D.C.: NRC, 1964); included in Penick et al., *op. cit.*, 1965 (19), p. 33.
52. Some transfers were made in the early seventies, forced by the Mansfield amendment, but this has not significantly affected the pattern in the longer term (aside from the Nixon administration's exceptionally high allocations for cancer research through the NIH). Indeed, with the rise of SDI, the share of the military within university physical science research seemed set to rise further, and be more explicitly tied to defense missions.
53. Cited by NAS, *op. cit.*, 1964 (51); emphasis added.
54. Haberer, *op. cit.*, 1969 (47), pp. 195, 229, 273—274, 227, 224; A. H. Compton, *Atomic Quest: A Personal Narrative* (New York: Oxford University Press, 1956), pp. 216, 112—114, 118—119, 181; Leslie Groves, *Now It Can Be Told* (New York: Harper Bros., 1962), pp. 125, 129, 136. The 1946 victory of the atomic scientists' movement over the May-Johnson bill was snatched away by the passage of the Vandenberg amendment to the successful McMahon bill, providing for a Director of Military Applications and a Military Liaison Committee within the AEC, whose chief client and preponderant influence was, in any case, the War Department; see Haberer, *op. cit.*, 1969 (47), pp. 232—233. On Lawrence's support, see Heilbron et al., *op. cit.*, 1981 (39), pp. 48—49.
55. Haberer, *op. cit.*, 1969 (47), p. 227, citing the testimony of Bush (47), p. 843.
56. The "Bird Dogs," "The Evolution of the ONR," *Physics Today* **14** (August 1961), 35; Penick et al., *op. cit.*, 1965 (19), pp. 22—23.
57. Forman, *op. cit.*, 1987 (37); Kevles, *op. cit.*, 1979 (24), pp. 354—355.
58. E. Digby Baltzell, *The Protestant Establishment: Aristocracy and Caste in America* (New York: Vintage, 1966), pp. 210—211; Kevles, *op. cit.*, 1979 (24), pp. 211—215; Marcia Synnott, *The Half-Opened Door* (Westport, Conn.: Greenwood, 1979); Dan A. Oren, *Joining the Club: A History of Jews and Yale* (New Haven: Yale University Press, 1986); Charles Weiner, "A New Site for the Seminar . . . ," in Donald Fleming and Bernard Bailyn (eds.), *The Intellectual Migration* (Cambridge, Mass.: Harvard/Belknap Press, 1969), p. 216; P. K. Hoch, "The Reception of Central European Refugee Physicists of the 1930s: USSR, UK, USA," *Annals of Science* **40** (1983), 239 ff.
59. E.g., Bush, *op. cit.*, 1954 (47).
60. On the AIP, see above, n. 21.
61. Cited by Greenberg, *op. cit.*, 1969 (31), p. 178.
62. See esp. the exchange around the article by Louis N. Ridenour, "Military Support of American Science, A Danger?" *Bulletin of the Atomic Scientists* **3** (August 1947), 221—230; Morrison is quoted by Clayton R. Koppes, *JPL and the American Space Program: A History . . .* (New Haven: Yale University Press, 1982), pp. 258—259; Norbert Wiener, "A Scientist Rebels," *Atlantic Monthly* **179** (January 1947), 46 ff; James B. Conant, "An Old Man Looks Back: Science and the Federal Government, 1945—50," *Bulletin of the New York Academy of Medicine* **47** (November 1971), 1248—1251. The fullest discussion of this period is David Caute, *The Great Fear* (New York: Simon and Schuster, 1978), esp. chaps. 13 and 25; Martin D. Kamen, *Radiant Science, Dark Politics: A Memoir of the Nuclear Age* (Berkeley: University of California Press, 1985).
63. Sapolsky, *op. cit.*, 1979 (34), pp. 385—386, 395, who cites Compton's comments

from "Transcript of the Second Meeting of the Naval Research Advisory Committee" (Washington, D.C., January 15, 1947), p. 47; U.S. Army, *Final Report by the Advisory Committee on Contractual and Administrative Procedures for the Department of the Army* (Washington, D.C.: U.S. Army, October 15, 1945).

64. DuBridge, *op. cit.*, 1949 (25), pp. 279—280. See also Walter Gellhorn, *Security, Loyalty and Science* (Ithaca, N.Y.: Cornell University Press, 1950).
65. L. V. Berkner, "University Research and Government support," *Physics Today* **17** (January 1954), 10—17; cited by Forman, *op. cit.*, 1987 (37).
66. This question is also addressed in slightly different form by M. S. Sherry, who claims that the science elite's "fastidiousness about their autonomy from controls obscured from them the changing role played by their profession" (M. S. Sherry, *Preparing for the Next War: American Plans for Post-War Defense, 1941—45* [New Haven: Yale University Press, 1977], p. 158).
67. E. U. Condon, "Reminiscences of a Life in and out of Quantum Mechanics," in *International Symposium on Atomic, Molecular, Solid State Theory and Quantum Biology VII*, ed. P. O. Lowdin (1973), pp. 7—22. I am indebted to Mrs. Natalie Gurney-Taylor for making available information on her late husband, including the full text of his security hearings.
68. Forman, *op. cit.*, 1987 (37); Caute, *op. cit.*, 1978 (62), pp. 479—484. Other prominent mathematicians and physicists hounded in this period include Dirk Struik, Melba Phillips, and Wendell Furry. For a Canadian example see Leopold Infeld, *Why I Left Canada*, ed. Lewis Pyenson (Montreal: McGill/Queens Press, 1981).
69. Haberer, *op. cit.*, 1969 (47), chaps. 10—11; Barton J. Bernstein, "In the Matter of J. Robert Oppenheimer," *Historical Studies in the Physical Sciences* **12** (1982), 195—252.
70. E.g., Seymour Melman, "Who Needs a War Economy?" *The Nation* **215** (November 20, 1972), 487 ff; Michael Kidron, *Western Capitalism since the War* (Harmondsworth, Middlesex: Penguin, 1970); Lewis Mumford, *The Pentagon of Power* (London: Secker and Warburg, 1971).
71. Sapolsky, *op. cit.*, 1979 (34), p. 389.

PART II

THE MILITARY AND TECHNOLOGICAL DEVELOPMENT

THE COMPUTER: A CASE STUDY OF SUPPORT BY GOVERNMENT, ESPECIALLY THE MILITARY, OF A NEW SCIENCE AND TECHNOLOGY

I. BERNARD COHEN
Harvard University

Introduction

The electronic, digital, stored-program computer came into being with massive government support — largely, but not exclusively from millitary agencies. Wartime needs provided an enormous stimulus to the design and development of new machines for calculation or computing. It is anyone's guess whether or not the computer as we know it today would have come into existence without this financial backing, but we may be certain that the pace of development would have been much slower had there not been these funds from government. From World War II until well into the 1950s, development funds for the giant steps of advancement that the computer required exceeded by far the capacities of industry, of universities, and of private foundations (1).

The first stage of computer history has some remarkable features that differentiate it from conventional history of science and technology (2). In both Germany and America in the 1940s, as we shall see, those who were developing the first computers or proto-computers were "new men" in the arena of science and technology, outsiders who had not been active or prominent members of any traditional "old boy" network. J. V. Atanasoff received some nominal support from Iowa State College, but his work was out of the mainstream of computer technology and was not very much known in wider circles. George Stibitz was a member of a large and well established research and development organization, the Bell Telephone Laboratories, but in-

E. Mendelsohn, M. R. Smith and P. Weingart (eds.), Science, Technology and the Military, Volume XII, 1988, 119–154.

novations in computing or machine calculation were at most a minor peripheral concern of his fellow scientists and engineers. Howard Aiken was a real outsider and upstart, only a graduate student in a field of science, physics, far removed from any concern with inventing new machines for numerical calculation. He related that he had been told by the permanent tenured members of the Physics Department at Harvard that they had no interest in his proposed machine and would not give it any support, and he maintained that Harvard's President Conant had even told him that he would have no future at Harvard if he continued to work on computing machines rather than doing more traditional work in electron physics (3). J. P. Eckert and John Mauchly, working on ENIAC in the 1940s, were considered to be outsiders, whose project seemed to have no interest to those who were concerned to advance methods of practical computing (4). And in Germany, Konrad Zuse was a real "loner," an outsider with no academic affiliation and no industrial support. His pioneering work in designing and building his first machines made use of scraps and second-hand spare parts.

In England, the situation was somewhat different. The first group of supercalculators or computers was designed and constructed by a group at Bletchley Park, London, as a direct response to wartime needs for encription and decoding. The computers they built were electronic and very powerful. The full history of this operation was long withheld from the public knowledge by the Official Secrets Act. (It may be noted that there is still no full public information concerning the machines and programs used in the United States during World War II for encription and decoding. The success of this effort — for example, in breaking the Japanese Secret Naval Code — indicates that some kind of powerful machines must have been in use [5].)

Some General Features of Computer History

The history of the computer may be best understood by considering three main aspects: technology, architecture, and computer art and science. The rubric of technology includes the invention and application of such new or improved functional elements as memory devices — mercury delay lines, Williams (cathode-ray) tubes, magnetic cores —

transistors, integrated circuits, chips, and solid logic. Architecture is the name given to the "physical or hardware structure" of computer systems, their individual properties, and the various ways in which they may be linked. Originally, this field was limited to actual machine design, as a subsection of electrical and electronic engineering — that is, the design of electrical and electronic circuits and their "organization into a computer system"; hence the great importance in early computer architecture of the subject of switching theory. The art and science of computing was traditionally concerned with the theory and practice of operating systems, with the development of software and with general questions of computability and automata. Later, computer architecture came to embrace both basic machine design and elements of computer science, and even aspects of operations research and solid-state physics, along with the more traditional electrical and electronic engineering and some software considerations (6).

In considering the origins and development of the computer, we find that each of these three aspects shows dual, complementary phases of development: a logic of internal innovation and reaction to the introduction of new technological elements, and also a direct response to external, market-driven demands. In both of these phases we shall see that government has played a crucial role.

In the United States the early design and construction of computers took place primarily in universities with funds supplied by the military agencies, the Navy, the Army, and the Air Force. Later on, financial support was available to some extent from other agencies, such as the Atomic Energy Commission, the National Security Administration, NASA, the Bureau of Standards, and the Weather Bureau. The first commercially produced computer, UNIVAC I, begun in 1948 and completed in 1951, was delivered to a civilian agency, the U.S. Bureau of the Census. It was apparent, however, that the large-scale funding needed to develop computers would come primarily from military or military-related sources.

The early computers that did not go to government installations, or that did not remain in universities where there was government support, were used primarily in engineering or applied science research and development (notably in the aviation industry) on projects that had government backing; they were also used on national security problems, such as code-breaking, encription, and decoding. The first commercially planned computer appears to have been the BINAC, a small binary machine, designed by Eckert and Mauchly (who founded a commercial company in 1947, after leaving the Moore School of the University of Pennsylvania) for the Northrop Aircraft Corporation (7). The BINAC (Binary Automatic Computer) became operational in August 1949, the first stored-program electronic digital computer to be completed in the United States.

The government and government-sponsored organizations were not only consumers in the sense of buying the machines as finished products: in the 1940s and 1950s they also set forth specific needs or demands for which machines were to be constructed. Thus Howard Aiken, at the Harvard Computation Laboratory, designed and produced two machines for the Navy's Weapons Testing Laboratory at Dahlgren and a final machine for the Air Force (8).

In the 1950s a significant change occurred in the computer field: by the end of the decade, computer design and construction had shifted from universities to commercial manufacturers (9). We shall see, below, however, that even after this shift from universities to industry, the high cost of computer development in the United States continued to be supported by government, notably the military and military-related agencies, through direct subvention to manufacturers.

Some Military Components of Computer Development

There is a wartime link between scientists and engineers and later computer development that is often not fully recognized: in England and in America, many computer scientists and engineers received fundamental experience and training in electronics as a result of working with radar. Radar experience was certainly an important factor in the career of Maurice Wilkes, whose contributions will be discussed below. This aspect of the subject has been beautifully put by Wilkes. He writes of a new breed of men who, like himself,

thought of ourselves as electronics men, not as electrical engineers. We were used to wide bands with short pulses and we saw the possibility of achieving very high speeds with elegant economy of equipment by these techniques. I think there was something of an age gap about it. Very few people more than five years older than I was somehow — what shall I say? — had "green fingers" in electronic circuits. It was a young man's game (10).

Another area in which the effect of the military establishment can be discerned is in the conferences and training programs organized in the years after World War II. We must keep in mind, as Herman H. Goldstine has reminded us, that in the post-World War II years there was a "great awareness of the scientific importance of computers" among "the university and government community," and that "the earliest and boldest developments [were] . . . in general in the university rather than in the industrial world." The "seminal ideas developed largely in the university world, with the help of far-sighted government support":

[The] basic notions which underlay the computer revolution . . . were disseminated to the world by means of a number of large scientific meetings and by reports written for the government and given wide circulation by various governmental agencies. Notable among the meetings were several organized by Aiken at Harvard, which were widely attended. Perhaps the most noteworthy of the reports was a series of papers that emanated from the Institute for Advanced Study starting in mid-1946 . . . (11).

In this connection, we may observe that the two conferences organized by Aiken "on large-scale digital calculating machinery" were (as the title page of the *Proceedings* explicitly declared) "jointly sponsored by the Navy Department Bureau of Ordnance and Harvard University" (12).

In retrospect we may gauge the importance of the first Harvard symposium (1947) by the number of "early computer luminaries" among the participants: they included Howard Aiken, Wallace Eckert, Robert Everett, Jay Forrester, Herman Goldstine, Richard Hamming, Grace Hopper, John Mauchly, George Stibitz, Alan Turing, and Norbert Wiener (13). Yet another significant gathering took place at the Moore School, the home of ENIAC, in 1947 and 1948. The lecturers included Stibitz, Mauchly, D. H. Lehmer, Douglas R. Hartree, Goldstine, Arthur W. Burks, J. Presper Eckert, Aiken, J. H. Curtis, John von Neumann, Jan Rajchman, and Perry O. Crawford — a veritable *Who's Who* of computing in the late 1940s. Among the students were Everett, Claude Shannon, and Wilkes. This notable event was funded jointly by the U.S. Navy Office of Research and Inventions and the U.S. Army Ordnance Department (14). In England, in 1949, Maurice Wilkes organized a conference on High-Speed Automatic Calculating Machines at the University Mathematical Laboratory, the computer center at Cambridge University. This conference and the publication of the resulting volume of proceedings were supported by the Ministry of Supply (15).

A main focus for developing machine-aided computation and computer facilities in England was the activity and interest of scientists at the Ministry of Supply. Eventually there was a "practical outcome" in the establishment of a Mathematics Division in the National Physical Laboratory (NPL) at Teddington, Middlesex, which was fortunate in enlisting the services of Alan Turing. The latter soon had a proposal ready to build an electronic stored-program computer (February 1946). The director of NPL, Sir Charles Darwin, sought to enlist the cooperation of the Post Office staff, which included several veterans of the Colossus project at Bletchley Park (see below), but they proved to be too occupied with telephone construction to be able to go ahead on any computer projects (16). The Post Office did, however, through the Telecommunications Research Establishment (TRE) at Malvern, become "involved [in] the construction of a computer called MOSAIC for defense applications." MOSAIC (Ministry of Supply Automatic Integrator and Computer) was developed by a Post Office Team during 1947—54 for "tracking and telemetry problems associated with guided weapons," notably "processing radar tracking data" (17). As late as 1980, Lavington reported that parts of the MOSAIC project "are still secret" (18).

It may thus be seen that the history of the computer has been linked in many different ways to the support of military agencies and to military needs and experience. In exploring some aspects of the relations between the computer and the needs of defense or of the armed forces, in what follows, I shall limit my presentation primarily to the history of this topic in the United States.

Background to the Computer Age

It is a matter of record that from the nineteenth century to the mid

1930s, the chief activities — other than accounting, billing, and inventorying — requiring extensive numerical computing were the production of astronomical (and navigational) tables and of actuarial tables. While actuarial tables have no obvious connection with military affairs, the production of astronomical tables (ephemerides) has for three centuries been the direct responsibility of the navy in most (if not all) countries of the world, because such tables have had a primary use in navigation. The need for these two types of tables spurred the invention of new machines. Thus Charles Babbage's designs for his two supercalculators, the Difference Engine of the 1820s and '30s and the Analytical Engine of the 1830s, were motivated in the first instance by the existence of many errors in the astronomical tables produced by the Admiralty, and in other tables. A modified Babbage Difference Engine — designed and constructed by Georg and Edvard Scheutz — was actually used in 1864 in England for computing William Farr's celebrated *English Life Tables* and in the United States for computing astronomical tables (19).

Having introduced Babbage, I should mention that he received rather large-scale support from the British government for the construction of his Difference Engine. Hence, at the very beginning of the era of modern computing by machine, the major source of financial support was the government (20). At about the same time, Charles Xavier Thomas de Colmar invented, perfected, and manufactured and sold the first commercially successful digital calculating machine, the "Arithmomètre" or "Arithmometer." His customers included the French and British armed forces in addition to banks and insurance companies.

In the twentieth century there was a growing demand for engineering tables. World War I sharpened the awareness of a need for accurate firing tables for artillery. However, most scientists and engineers of the 1920s and 1930s had no use for the many decimal places that would be made possible by an advanced automatic digital calculator; they were satisfied with the three-place accuracy provided by Vannevar Bush's Differential Analyzer and other analog devices (21). (Recall that a good slide rule gives three significant figures.) In the 1930s, the great potential of business machines for scientific calculation was only just being realized. The real pioneer in this area was L. J. Comrie, a New Zealander established in London (22). Comrie, who had been superintendent of the British Nautical Almanac office, established his own

Scientific Computing Service. In particular, he recognized that commercially produced business machines had the capability of being linked together so that the output of one might become the input of another, thus functioning as Babbage's Difference Engine was intended to do (23).

A fruitful line of development in the 1930s made use of punched-card machines. These had been developed in linear descent from the original data-processing punched-card machines invented by Herman Hollerith to handle the data accumulated in the 1890 census (24). Hollerith's company was one of the three that were combined in 1911 to form the Computing-Tabulating-Recording Company (CTR) – out of Hollerith's Tabulationing Machine Company, the Computing Scale Company, and the International Time Recording Company; in 1924, under Thomas J. Watson's leadership, CTR became the International Business Machines Corporation (25). In 1928, IBM funded a center at (or associated with) Columbia University for advanced digital calculations. This installation, known as the Thomas J. Watson Computing Bureau, had originally been conceived as a statistical bureau for educational testing (pioneered by Benjamin D. Wood). Under the direction of Wallace Eckert, a distinguished astronomer, it pioneered in the use of punched-card equipment for scientific problems. A notable example was the computation of tables of the motions of the moon (26). At this time, in England, punched-card tabulators were being used in a similar way for scientific computation by L. J. Comrie.

It may be observed that punched-card machines are essentially cumulators. Hence, since multiplication is a form of controlled addition, and since addition is successive cumulation, a punched-card statistical or business machine may easily be directed to perform additions and multiplications. In the IBM punched-card business machines, subtraction was performed as addition, using complement arithmetic (27).

In the 1920s and 1930s, a "computer" was generally a man or woman operating a desk calculator. These were often still driven by hand, turning a crank, although increasingly they were becoming electrically powered. For integration, the solution of differential equations, hand planimeters were used (28) and, eventually, the Vannevar Bush Differential Analyzer (29). Other types of analog computers were designed for special purposes — e.g., the Kelvin tide meter and the Michelson harmonic analyzer. Much of the calculating world was satisfied with the results of these various systems. For engineering purposes the three-place limit of the Bush Differential Analyzer posed no problems. But in a number of different places, both in America and on the Continent, there were independent developments that show us the rise of an internal demand for greater calculating power. At the same time, in the late 1930s, Turing developed the abstract concept and theory of a computing machine and, a little later, Claude Shannon conceived his own view of communication theory.

The First Computers or Proto-Computers

The great figures associated with giant calculators or computing machines in the late 1930s were not directly concerned with problems arising from practical engineering needs or with military affairs. George Stibitz started out by amusing himself with the analogs of relay circuits and binary arithmetical operations; he was then led to design and build his first practical machine in order to facilitate computations with so-called complex numbers. As he himself wrote, in a report of 1940: "For some years . . . need has been felt for a computing machine which would relieve the operator of the numerous details involved in complex computations" (30). Actually, Stibitz had begun to experiment with relays in connection with binary arithmetic as early as 1937, when he was exploring the common features of binary arithmetic and electric circuits containing relays. He constructed a model in his kitchen using a block of wood, strips cut off a tin can, batteries, relays, and two flashlight bulbs (31). Some time after he had shown his toy (called "model K" — for kitchen) to his colleagues at the Bell Labs, he was urged by T. C. Fry — a distinguished applied mathematician at Bell — to develop it into a practical device. He accordingly began to design a calculator for complex numbers for practical use in designing filter networks. The first machine (developed at Bell Labs by Stibitz and S. B. Williams) was completed in October 1939 and put into regular use in January 1940; it was known as the "Complex Number Computer" (later shortened to "Complex Computer"). It could, according to Stibitz, speedily and easily add, subtract, multiply, and divide complex numbers: "All it requires of the operator is that she type the problem correctly" (32).

The machine was publicly demonstrated in September 1940, at a meeting of the American Mathematical Society at Dartmouth College. A teletypewriter was installed that — by remote control — operated the machine in the Bell Labs on West Street in New York City. This was "the first machine to service more than one terminal" and "also the first machine to be used from a remote location" (33). Model I, as the machine was later called, was thus originated and developed in an industrial milieu, independent of government and wholly unrelated to military problems or demands. Stibitz's machine was put into operation almost at once and continued to function at the Bell Labs for many years. For Bell it was the end of the road, not the start of a new venture into automatic digital computational machines. But the onset of war changed the situation drastically (34). This effect has been described by Stibitz as follows:

> The advent of the War almost simultaneously broke down the other barrier to progress. While the demand for rapid and inexpensive computation had been increasing for years before the war, the increase was, as we look back at it now, slow and insignificant in comparison to that which the war engendered. Hind sight is notoriously superior to foresight, and it seems strange, on looking back, that we did not see immediately the urgency with which computation after computation would be needed in the laying of the scientific ground-work for the conflict. New military devices, unsolved problems of tactics, unfamiliar subjects, — they all demanded mathematical treatment if possible. And mathematical treatment meant finding an answer immediately, not after a leisurely program of trying out one classical method after another. Delays meant not merely money but lives lost (35).

Stibitz himself went on to a wartime administrative post in the National Defense Research Committee of OSRD, in which he was responsible for Bell Labs' undertaking to develop two more machines. The first was intended to simulate an automatic aiming device for anti-aircraft guns; later called the Model II Relay Calculator, it became operational in September 1943 and eventually was used for other computational problems as well (36). The companion machine was designed in 1942, even before the Model II Relay Calculator had been finished; known as the Ballistic Computer, this machine served a function similar to, but somewhat more complex than, that of its predecessor (37).

The case of John V. Atanasoff once again shows how the initial impetus for digital computation was wholly unrelated to military problems or military support. In the mid-1930s at Iowa State College (now Iowa State University) Atanasoff was motivated by a concern with applied mathematics (systems of linear equations in engineering problems) to go beyond the limits of analog calculation and to explore the possibilities of digital calculating. (Atanasoff has claimed that he invented the term "analog computer.") A major problem for him was "the solution of systems of linear simultaneous algebraic equations" (38). He was joined by Clifford Berry in 1939, after receiving a grant for building the machine (in the amount of $650 from the Iowa State College Research Council), later supplemented (spring of 1940) by a further grant of $700 (39). The machine, intended to solve up to thirty simultaneous linear equations, was to be equipped with a binary card-punch and reader developed by Berry. A "prototype computing element" was demonstrated in autumn 1939, but the large-scale machine was never completed. The project was abandoned when Atanasoff left Iowa State to join the staff of the Naval Ordnance Laboratory in 1942, at the same time that Berry joined an engineering company in California.

The Atanasoff-Berry computer (or ABC) has credit for a number of "firsts": it was certainly a pioneering calculator or computer to use vacuum tubes, and it ("or its 1939 prototype") has been held to be "the first machine to incorporate a regenerative memory." The ABC achieved real fame during the famous 1971—73 court case, *Honeywell v. Sperry Rand*, in which Honeywell contested the ENIAC patent, then assigned to Sperry Rand (ENIAC will be discussed below). On 19 October 1973 the presiding trial judge handed down his verdict in favor of Honeywell, declaring that "Eckert and Mauchly did not themselves first invent the automatic electronic digital computer but instead derived the subject matter from one Dr. John Vincent Atanasoff" (40). From that day till now there has been a running debate on Atanasoff's claims and the possible indebtedness of Mauchly to Atanasoff (41).

A third significant American development, associated with Howard H. Aiken, was also without any connection to military problems of computing. Aiken turned to the design of a supercalculator in the mid-

1930s, while still a graduate student at Harvard. He wrote up a final prospectus of the architecture of his proposed machine on 4 November 1937, long before there was any military situation that might demand such a machine (42). The research undertaken by Aiken for his doctoral dissertation, on a problem in electron physics associated with vacuum tubes, had given rise to sets of differential equations that he could not solve in the ordinary way. This turned his inventive mind to machine solutions, to the invention of a machine that could be programmed to produce automatic sequences of operations. IBM eventually agreed to construct such a machine, but it was not completed and delivered to Harvard to become operational until August 1944 (43). By this time, America was deep in the war and Aiken was on active duty in Virginia as a naval officer. The new computer, Mark I, was immediately put into service as a Navy instrument, operated under the Bureau of Ships of the Navy, and Commander Aiken was sent back to Cambridge to be in charge. The staff consisted of a number of mathematically trained naval officers, plus the usual complement of "ratings."

Although Mark I thus begun its active career as part of the Navy, it was not constructed with funds provided by the Department of the Navy. Rather, it was built entirely by IBM with its own resources (44). Begun in 1937, two years before the outbreak of war in Europe, and four years before America's active involvement, Mark I — in its original conception — was in no way directly related to the military. As a matter of fact, in discussing the need for such a machine, in his prospectus, Aiken stressed applications in the physical sciences, pointing out that in theoretical physics there existed present "problems beyond our ability to solve, not because of theoretical difficulties, but because of insufficient means of mechanical computation." The nearest he came to any kind of practical problem was the mention of "the study of the ionosphere," where, he said, "the mathematical expressions required to represent the phenomena are too long and complicated to write in several lines across a printed page" but represent the "type of research" on which "rests the future of radio communication and television" (45).

An examination of the work produced by Mark I during the last two years of the war shows that a machine of this type could provide solutions to problems of immediately pressing practical issues and valuable tools of more general or theoretical interest or usefulness, which were nevertheless related to ultimate military applications. Thus Mark I worked on a problem of implosion, set by John von Neumann as part of the work on the bomb at Los Alamos, and on ray tracing in the design of lenses for aircraft reconnaissance photography (46). At the same time, however, the machine produced tables of Henkel functions and Bessl functions, which while of direct use in solving some practical military oriented problems were also of importance in wide areas of applied mathematics (47). It is notable that Mark I never produced ballistic tables or other direct aids for military gunnery, but did yield results of use in protecting ships from magnetic mines. In short, Mark I was conceived and planned in an environment far removed from the stress and demands of war — but it was sufficiently flexible that it could easily be programmed to solve military problems when they arose.

In Germany the beginnings of the modern computer were were also completely independent of wartime needs or problems of military origin. The pioneer in the area of computers in Germany was Konrad Zuse. Zuse was graduated from the Technische Hochschule in Charlottenberg-Berlin in 1935 and started his career as a stress analyst for the Henschel Aircraft Company in Berlin. Even before graduation, he had begun to think about a universal calculator that would use binary arithmetic; so when his work at Henschel led to a dreary sequence of setting up and subsequently solving systems of simultaneous linear equations, Zuse — like Stibitz, Atanasoff, and Aiken — decided that the calculations could and should be mechanized. At the age of twenty-six (in 1936), he started to build his first machine (48).

Zuse's first two calculators were actually built in the living room of his parents' apartment with the help of friends. The first of these, the Z1, was a binary machine, controlled by punched tape; it had a mechanical memory consisting of a set of thin slotted metal plates, a thousand in number, in which the position of a pin in the slot would indicate a 0 or a 1. Although it was never fully operational, the Z1 convinced Zuse that he was on the right track. He then began collaboration with a gifted electrical engineer, Helmut Schreyer, with whom he produced the Z2. This machine was in many ways similar to the machines of Stibitz and Aiken in that it used telephone relays. It evidently functioned rather well, whereupon Schreyer conceived the idea of building a yet more powerful and rapid machine using vacuum tubes. Before the Z2 had been completed, however, Zuse was drafted into the German army following Germany's invasion of Poland in 1939. One of his sponsors eventually got Zuse freed from the German army so as "to complete" his "work on an important invention." Zuse returned to Henschel while Schreyer (not drafted into the army) completed his doctorate, his thesis subject being the use of vacuum tubes as digital switches.

It should thus be clear that the beginning of Zuse's involvement in computers was similar to Stibitz's, with no direct support from the military. Also like Stibitz, Zuse was concerned with a problem in applied mathematics related to work in engineering or technology. Of course, it can be argued that Zuse's work was indirectly related to military problems, since the simultaneous linear equations he had to solve arose in connection with aircraft design, and hence with military aviation technology. But this is different from direct military support of his invention or even a military motivation for making the invention in the first place. In fact, the military were not particularly interested in Zuse's invention; Zuse paid all the costs of the Z1 and Z2 out of his personal funds.

In 1942, Zuse's collaborator Schreyer proposed to the German Army Command that they authorize him to construct an electronic calculator, containing about 1,500 vacuum tubes, and able to execute some 10,000 operations per second. Schreyer's project received no support. The army's high command apparently was unwilling to fund any project that did not seem to hold promise of an immediate contribution to the war effort. Supremely confident that the war could not last more than a couple of years, the High Command was not interested in any long-term project. At the same time, Zuse was making application to the Aerodynamics Research Institute for support of his proposed computer. We are told that this group had no interest whatsoever in a *general-purpose* calculator, but was willing to support Zuse if he could help them with a single, but major, computational problem. The calculation in question, needed so urgently, had to do with the wing flutter of an aircraft. Zuse relates in his autobiography that he told them he could design a *special-purpose* calculator that could automatically solve the equations, and he was granted permission to build a prototype. What he intended, however, was to devise what was in effect a general-purpose machine that could solve a variety of problems, of which wing flutter would be but one. Zuse, commissioned by the Aerodynamics Research Institute, set up a small company; his Z3 was completed by the end of 1941, to become what was probably "the first operational general-purpose program-controlled calculator" (49).

We have Zuse's word for it that the Z3 performed its operations very well (50). But those in charge at the Aerodynamics Research Institute seem to have decided that their needs would be better taken care of with simpler, special-purpose machines. Zuse thereupon designed for them (and oversaw the construction of) two calculators with the limited function of being able to analyze the wing flutter of flying bombs (51). The machines were so wired that they could automatically perform a series of calculations as the bombs came off the assembly line, yielding the result of the amount of adjustment of the wings needed on each one individually, so that flutter could be minimized if not eliminated. Zuse tells us that before his calculators were introduced, there had been thirty women using mechanical desk calculators to solve these problems; these women were now transferred to other jobs.

The Z3 was completely destroyed in an air attack on Berlin in April 1945. Zuse, in the meanwhile, had begun to build a larger version, the Z4. As this neared completion, it had to be moved several times in Berlin when the structures housing it were nearly destroyed by Allied bombing. He and his assistants eventually arranged to have the Z4 transported to Goettingen, and later on to a village in the Bavarian Alps, where he and the machine fell into the hands of the American Army. Eventually, the Z4 was installed in Zurich, where it did important service during the postwar years (52).

The War Years

I have already referred to some of the prewar developments that became significant for the war effort in Germany and in America, in relation to the activities of Zuse, Stibitz, and Aiken. At Harvard, as a Navy unit, Aiken and his group undertook to build a supercalculator for the Navy Proving Ground at Dahlgren. Like Mark I, this machine (Mark II), begun in 1944, was operated by electromechanical relays. Aiken was not satisfied with the performance of standard relays and designed better ones, which were specially manufactured for him by the Autocall Company. Mark II had an interesting architecture in that it was really two machines in one: each of the pair could be operated separately and independently, or the two could be tied together in either serial or parallel connection (53).

In the prewar years IBM had sponsored the laboratory at Columbia University where Wallace Eckert used IBM punched-card equipment for such problems as calculations of the orbital motion of the moon (54). IBM's product line then (starting from 1935) included a "multipying punch," the IBM 601, which contained a relay-based arithmetic unit that required about one second to multiply two numbers. IBM produced about 1,500 of these 601s, which, "because of their ready availability, formed the backbone of most of the scientific and commercial calculation until the advent of the electronic computer" (55). Such machines not only provided more speed of calculation than an electrically driven desk calculator but, "more importantly," also allowed for increased accuracy "because of the needs for fewer human interventions in the process" (56).

During World War II IBM designed and produced a special machine for the U.S. Army, to serve in the production of ballistic tables for the new artillery. Called the IBM Pluggable Sequence Relay Calculator (PSRC), it used standard IBM technology, based on relays. The control technique was to have the operations directed through a series of special relays connected to plug-board wires, with the data fed in through punched cards. Although not notably innovative in terms of technology, the PSRC was much faster than the standard IBM 602; it could readily perform addition, subtraction, multiplication, and division, and could also calculate square roots (57). Two of the PSRCs produced in 1944—45 went to the Aberdeen Proving Ground. Later, three more were made for the Navy's use at Dahlgren and another two for the T. J. Watson Scientific Computing Laboratory at Columbia. The

PSRC was never part of the IBM product line and it is "generally considered as one of the special-purpose war-time calculators rather than just a modification of their multiplying punch" (58).

The most significant wartime development in computing was ENIAC (Electronic Numerical Integrator and Computer), designed and produced at the Moore School of the University of Pennsylvania under contract to the U.S. Army, specifically for the Army Ordnance Department Ballistic Research Laboratory at the Aberdeen Proving Ground in Maryland. (Before the onset of World War II, the Moore School had cooperated with the Ballistic Laboratory in constructing two Differential Analyzers for use in the determination of ballistic trajectories. Vannevar Bush, the inventor of the Differential Analyzer, had helped in this venture, and the two machines became operational in 1934 [59]. In this connection it is not amiss to record that the Differential Analyzer became the "workhorse" calculator during the war years, a fact that we are apt to ignore in our concentration on digital machines to the exclusion of analog devices [60].)

Many heroic figures played a significant role in the planning and development of ENIAC. Herman Goldstine, a mathematics professor serving as a lieutenant in the Army, had the responsibility of seeking whatever devices or ideas might speed up the production of artillery or ballistic tables (61). He provided the liaison between the Army and the Moore School and spear-headed the support for the new machine. His primary contact at the Moore School was John Brainerd, who became the administrative head of the project. The two men responsible for the new idea and its implementaion were John Mauchly and J. Presper Eckert, the former a physicist, the latter an electronics engineer. Others of importance were Arthur Burks, a mathematician and logician, Joseph Chadakar, and Kite Sharpless. Historical analysts generally give primary credit to Eckert and Mauchly.

There are many roots of ENIAC, but a convenient starting point is a short paper by Mauchly in August 1942 on "The Use of High Speed Vacuum Tube Devices for Calculating" (62). It is a matter of record that Mauchly had met Atanasoff in December 1940, at a meeting of the AAAS, after which he went to Iowa to see the work being done on the ABC. In the final production of ENIAC, Mauchly may be considered responsible for the "conceptual design" or general architecture, while Eckert had the enormous responsibility of designing the "individual circuits." The degree of their achievement can be gauged by the fact that ENIAC would be several orders of magnitude larger than any electronic device ever conceived or manufactured, and it would have to be reliable.

ENIAC marked a turning point in the history of computers. It provided the first demonstration that giant supercalculators could be built with speedy vacuum tubes rather than slow relays and that the results would be reliable. Just as Aiken's Mark I demonstrated that large-scale, automatic, sequenced calculation would give reliable results, so ENIAC moved a giant step into the future by showing the power of electronic circuits for computers. It became operational at the Moore School in the spring of 1945 and was later shipped to Aberdeen. By the time it was taken out of service on 2 October 1955, it had probably done "more arithmetic than had been done by the whole human race prior to 1945" (63).

The ENIAC group was fortunate in eliciting the interest of John von Neumann, one of the foremost mathematicians of the world (64). He crystallized an important new concept, of which major features had already been under discussion in the Moore School group: the "stored program." It is of no purpose here to enter into the hotly debated question of who invented the stored program. There seems to be no doubt that the idea came into being at the Moore School among the ENIAC group — chiefly Eckert and Mauchly, and Goldstine. There is equally no doubt that what made the stored program an actual part of computers was the celebrated draft report on the EDVAC, written by von Neumann under the date of 39 June 1945, from the Moore School, under contract with the United States Army Ordnance Department (65). Here was set forth "the first documented discussion" of "the advantages of using just one large internal memory, in which instructions as well as data could be held" (66). The proposed machine was to be known as EDVAC.

The name EDVAC was an acronym for "Electronic Discrete Variable Arithmetic Computer." This proposed machine, using the stored-program idea, and being electronic and digital, incorporates all the essential defining ideas of the modern computer. It was, not, however, the first such machine to be built. It was preceded by a "small experimental machine" built at the University of Manchester in 1948; its primary purpose was not to test the stored-program concept, but rather to test the new Williams tube (cathode-ray tube) system of storage (67). The director of the project was Prof. M. H. A. Newman, who "had been in charge of the secret special-purpose computers built at Bletchley" (which shall be discussed shortly) (68). The first electronic, digital, stored-program machine to be designed and built and put into regular, continuous, full-scale operation was EDSAC (Electronic Data Storage Automatic Calculator), at Cambridge University. The leader of the group that produced EDSAC was Maurice V. Wilkes, who had attended the famous Moore School where the design of EDVAC had been a major subject of discussion (69). When EDSAC was put into operation in May 1949, the computer age had definitely begun.

These American wartime activities may be compared and contrasted with what was going on in Britain – chiefly at Bletchley Park, just outside London, the center of Britain's Code and Cipher School. Here there was developed a machine that used both electronic and relay components to help in decoding German military information sent out with a special encoding device known as "Enigma." This first machine was known as the "Heath Robinson," named after the British cartoonist who was famous for his sketches of fantastic imaginary machines – a sort of super Rube Goldberg (70).

Many very talented British mathematicians, scientists, and engineers worked on this wartime project, among them Alan M. Turing, H. A. Newman, and I. J. Good. Only recently has full information about their activities come to light from behind the screen of the Official Secrets Act (71). We now know about the design, construction, and operation of the series of computers "Colossus") used for decoding, after the Robinson; their chief designer or engineer was Mr. T. H. Flowers of the General Post Office, who together with his group, had considerable experience and expertise in telephone switching. The first Flowers machine was completed and put into operation by December 1943. A second Colossus (Mark II) was rushed to completion by 1 June 1944, just five days before the Allies began the invasion of Europe.

The Colossus machines (there seem to have been ten of them altogether) were fully electronic (using 24,000 vacuum tubes), digital, and programmable. They were not computers to the degree that they did not "compute" numerical results, not did they print out numbers; but they "made a contribution to the development of computers in Britain by showing Turing, Newman, and others what electronics could do, and that knowledge turned their minds to computers immediately after the war" (72).

Despite their priority with respect to electronics and programming, the Colossus machines did not have any open and direct influence on the later developments of computers because they were shrouded in secrecy until the 1980s. Yet some of the scientists and engineers associated with Bletchley Park — e.g., Newman, Flowers, Turing — later applied their wartime experience to computer design. The influence of Zuse, similarly, was far less than the degree of his innovation would seem to have merited. He too constructed an early electronic machine. His Z3 was a general-purpose machine that incorporated two of the main functions of a modern computer: a memory (although not an electronic memory) and a capability of control by programming (although not a stored program). Unfortunately, as has been mentioned, the Z3 was destroyed in an air raid on Berlin. Zuse's early contributions did not become known, especially to the English-speaking world, until computer design and technology had advanced far beyond his seminal ideas. The main lines of development, as history unfolded, thus go back to Aiken and his Mark I, and especially to Eckert and Mauchly and ENIAC.

The Postwar Years: ONR and Computers

The history of the computer in the postwar years illustrates the theme of the importance of government in the dual roles mentioned at the beginning of this paper: as the source of funds for development, and as a primary consumer. The rapid and large-scale development of the computer and the computer industry in America could not have occurred as it did if the government had not provided large sums of money for these two purposes. For example, the recent technical history of IBM (produced in-house) shows the degree to which the expertise, the training, and the "know-how" for the advancements in computer science and technology so necessary for IBM's progress came about because of large-scale government funding for special-purpose machines. IBM's first entry into the computer field (where a computer is defined as an all-electronic machine, operated on a stored program) was the 701 computer. This machine, which began life under the name of "The Defense Calculator," originated during the time of the Korean War (73). Other examples of government subsidy leading to expertise that was later useful to IBM in its business are the NORC, or Naval Ordnance Research Calculator, which eventually replaced Aiken's Mark II and Mark III at Dahlgren (74); the work on ferrite magnetic-ore memories (produced for systems designed for SAGE [Semi-Automatic Ground Environment]); and project STRETCH (Space Technology Requirements Engineering Test of Component Hardware), leading to IBM's Type 7090, the first STRETCH-technology product (75).

In this connection it must be remembered that during the 1950s there was a parallel but independent development of machines for scientific (and engineering) and business uses. Only after a decade or so did it become widely practical to design and manufacture general-purpose computers that could function well in both domains. Since the largest sums of money were expended in engineering applications — in investigations supported by the government, or in government-supported industries such as aircraft manufacture — the development of computers for business was often at first an indirect offspring of the work in science and engineering.

Government funds did not come exclusively from military agencies. The first commercially produced computer was UNIVAC I, produced for the Bureau of the Census, a nonmilitary agency (76). Other non-military agencies associated with computer development included the Bureau of Standards, the National Security Agency, the Department of Agriculture, and the Department of Labor. In some cases there was both a civilian and a military component, as in space exploration, rockets, atomic energy, the various aspects of the aviation industry, and so on. Even so, those who have been studying the history of the computer are fully aware that military uses have always been an important component in the development and use of computers, not only for the branches of the armed services, but — as "Project Mac" and "ARPA network" remind us — even for the Pentagon directly (77).

One government agency in particular, the Office of Naval Research (ONR), was of notable significance in the development of computers during the all-important postwar years, marked by the coming of age of stored-program machines and the development of the magnetic-core memory (78).

ONR was established by an act of Congress in 1946, at a time when the wartime agencies — the National Defense Research Committee (NDRC) and the Office of Scientific Research and Development (OSRD) — had ceased their functions, and when there were seemingly endless delays and controversies concerning the establishment of a national science foundation (79). ONR was created specifically to fill the gap in the support of science, to ensure a continuity in the sponsorship of research, and to maintain the contacts between the Navy and the scientific community. The enabling act stressed the need "to plan, foster, and encourage scientific research" and declared the "paramount importance" of such research in relation to "the maintenance of future naval power and the preservation of national security." This goal required an institutionalized provision by the Navy of "the necessary services for conducting specialized and imaginative research." The latter phrase embodied the broad mandate for research support that, through ONR's Mathematical Sciences Division, was to prove vital in the development of the computer (80).

The Chief Scientist of the new organization was Alan T. Waterman, a former university professor of physics, who had held a major post in OSRD. Waterman brought a large part of his former staff with him. He was fortunate in getting Mina Rees, a former mathematics professor and wartime chief of the Applied Mathematics Branch of NDRC, to head ONR's Mathematical Sciences Division. A highly respected member of the academic community (she later became dean of the Graduate School of the City University of New York [CUNY]), she proved to be particularly sensitive to the needs of applied mathematics. ONR was a

Navy organization, headed by career naval officers, even though the policies were determined by the civilian scientists who set up the various programs and put them into effect. I have never encountered a mathematician or scientist who had contact with ONR during its early decade who was not enthusiastic about it as the optimal organization for the sponsorship of research (81).

The interests of ONR in applied mathematics emphasized such Navy-related topics as fluid dynamics, plasticity theory, theory of non-linear control mechanisms, and statistics and probability theory. There was a steady push for numerical analysis and various forms of "logistic research," embracing — inter alia — linear programming, game theory, and decision theory; these were related to ordinary "military problems of transportation, scheduling, warehousing, optimal resource allocation, and inventory management" (82). In the *Annual Report* of the Office of Naval Research for 1950, the need for a comprehensive research and development program in the computer area was spelled out in detail. The objectives were: "the development of high-performance computers, of improved computer components, and of simplified and relatively cheap equipment suitable for reproduction in university laboratories." Such "fast, high-performance computers" were needed, it was said, "not only for scientific computation," but also "for incorporation in information and control systems." In this statement we may note, as of 1950, the primacy of the needs of scientific (and, of course, engineering) problems in the design and use of computers.

In the ten-year period from 1946 to 1955 the machine developments supported by ONR included the original sponsorship of Whirlwind I (J. Forrester et al.) at the Massachusetts Institute of Technology and the IAS machine (John von Neumann et al.) at the Institute for Advanced Study. ONR directly sponsored the NAREC (Naval Research Electronic Computer) at George Washington University's Logistics Research Project, was associated with the development of NORC (see above), and supported research in numerical analysis at the National Bureau of Standards, using SEAC and SWAC (Standards Eastern/Western Automatic Computer). The new technologies proposed and tried for Whirlwind are like a roster of the cutting edge of the computer art — they include storage of data, new types of circuits, the use of magnetic tape, transistor circuitry, numerical methods, checking devices, computer reliability, and computer programming. By 1951, high-speed techniques developed for Whirlwind were making their way into "air-traffic control, industrial process control, insurance handling, inventory control, economic analysis, and scientific and engineering computations" (83).

The ONR-sponsored IAS machine, with its feature of parallel mode and asynchronous timing, bore fruit in its direct influence on many other computers, including the University of Illinois's ORDVAC and ILLIAC series, the Los Alamos MANIAC, Argonne's ADIVAC and GEORGE, Oak Ridge's ORACLE, and the Rand Corporation's JOHNNIAC. The list of foreign collateral descendants would include the BESK at Stockholm, the PERM in Munich, Australia's SILLIAC, and Israel's WEIZAC. It has been noted that the American machines just listed were designed for research purposes, "devoted primarily to scientific problems for national defense" (84). The major commercial machines of the late 1950s were also influenced by the IAS: IBM's 700 and 7000 series, and the UNIVAC 1100 series.

One feature of ONR's involvement with computer development is particularly noteworthy: the regularly published Surveys of Digital Computers. These were supplemented by a *Digital Computer Newsletter*, a means of keeping the whole computer community up to date on developments in the field. These documents show us graphically how the modern computer evolved. In the 1953 survey, well over half the machines listed were one-of-a-kind systems in universities or government installations, used primarily for scientific or engineering purposes; in this report, the emphasis is on university and government machines and on such special aspects as the different technological bases of memory. Two commercial machines are mentioned, the ERA 1101 and the UNIVAC I, along with the ONR-sponsored Whirlwind. In the 1954 report, stress on the machines has given way to a concern for applications: business use, machine translation from the Russian, election forecasting. Mention is made of the development and implications of transistors and magnetic-core memories.

The initial publication in this series was compiled by A.E. Smith and issued by the Office of Naval Research under the title, *Survey of Large-Scale Computers and Computer Projects* (1947, 1948). As the listed projects reached completion and new ones were begun, this work become outdated and was replaced by somewhat different work, undertaken for ONR by N. M. Blackman and published later in that year under the title, *Survey of Automatic Digital Computers*. The growth of computers in the interval may be gauged by the fact that the 1948 version was limited to a mere 12 pages (including planned devices or "projects"), whereas its 1953 successor contained 109 pages and was confined to completed, functioning machines. The extraordinary expansion of computer development can be seen even more dramatically by contrasting these publications with their successors. Martin H. Weik's *A Survey of Domestic Digital Computing Systems* (issued in 1955 by the Aberdeen Proving Ground as Report No. 971 of the Ballistics Research Laboratory) excluded all European machines and yet ran to 272 pages. Two years later, Weik produced his *A Second Survey* . . . (BRL Report No. 1010), which had expanded to 453 pages. *A Third Survey* . . . (BRL Report No. 1115) required 1131 pages for only those machines manufactured and placed in operation in the United States, comprising "two hundred twenty-two different electronic digital computing systems."

In the 1956 report, by Martin H. Weik, a change is noted: of the 44 electronic computer systems manufactured in the United States in that year, only 17 came from universities or government organizations, while 27 came from industry – chiefly UNIVAC, ERA, IBM, and CRC. The introduction of commercially made computers created a momentous change, signaled by the introduction of the IBM 650 in 1954. By the end of 1955 over a hundred of these relatively slow magnetic-drum machines were in service. What is significant is that users no longer had to design and build their own machine (or copy someone else's), since it was now more economical to purchase or rent a standard product. For the first time, a whole community of computer users were operating with systems that were identical or near-identical. As Herman Goldstine has observed, this was to have a most profound effect on the development of programming (85).

I shall conclude with two remarks that relate to the main themes of my presentation. First, the extremely high cost of the early dinosaurs of computing, the large mainframe machines, had the result that government support (primarily the military arm) was needed to provide — at least in major part — the funds for the initial research and development and, in many cases, for the eventual construction as well (86). Second, the mandate for such development represented a highly enlightened directorate for science and engineering within the Armed Forces of the United States. This latter aspect was stressed by John von Neumann in 1954 at the dedication of NORC, when he lauded the Navy and IBM for having been willing "to write specifications simply calling for the most advanced machine which is possible in the present state of the art" (87).

Supplement: The LEO Computers

The American developments, with large-scale government financing, may be contrasted with British computer history (88). I have already referred to Maurice Wilkes's work on EDSAC at Cambridge University. EDSAC was developed and produced with university funds and a small grant (89) from J. Lyons & Co. Lyons was in the wholesale food and catering business, operating some hundred and fifty tea shops, five hotels, restaurants, and an outdoor catering business (for such events as Wimbledon). The company was interested in computers because the management was concerned to improve and to automate or mechanize their statistical, payroll, and inventory work — in the late 1940s they employed a clerical staff of some two thousand. Under the leadership of two Cambridge graduates in mathematics — Raymond Thompson and John R. M. Simmons — Lyons sought to "rationalize clerical efficiency," rather than resort to punched cards. In 1947, Thompson and another executive went to America to consult with Herman Goldstine at Princeton, who put them in contact with Douglas Hartree and, through Hartree, Maurice Wilkes, then working on EDSAC at Cambridge, actually testing the "main prototype panels." It was then (November 1947) that "the Board decided to make a donation to Cambridge, and to lend the services of one of their engineers," Mr. Ernst Lenaerts, "for six months to Dr. Wilkes" (90). In September 1948 it was decided to build an EDSAC-like machine, and in January 1949

J. M. M. Pinkerton was hired as "electronic engineer" for this project (91). It was estimated that the proposed machine would entail a capital outlay of £100,000 and would save the company some £50,000 a year.

At first Lyons tried to have EDSAC replicated, but it was not possible to find any group ready and willing to make a copy. It was then decided to build a computer in-house, making changes and innovations as needed. By this time (May 1949) EDSAC was at last operational and passing its first test: to find prime numbers. It may be mentioned that a fundamental difference between a computer for scientific purposes (such as EDSAC) and a commercial machine was that the latter had much more input and output and less processing, because it had to deal with such problems as payrolls (with a variety of rates, taxes, and different deductions), and also, in this case, the problems of the ordering and supplying of some 40,000 different items to the 150 tea shops.

When the new computer was completed, it contained 5,000 vacuum tubes, had a stored program, and used half a ton of mercury for the mercury delay lines. The machine ran simple test programs in the spring of 1951 and was put into regular service for clerical work (operating at a rate of 700 instructions per second) in 1953. It was named LEO, an acronym for Lyons Electronic Office. It was so efficient that Lyons sold time and services to de Havilland Propellers Ltd. and the Ministry of Supply.

In mid-1954, Pinkerton set forth plans for a much improved version, LEO Mark II. In November 1954, a new company was launched by Lyons (Leo Computers Ltd.), which produced eleven models of LEO II, of which ten were sold and one kept for internal use. An improved version, LEO III, was made on a larger scale: more than one hundred LEO III machines were manufactured and sold to private companies, research organizations, and government installations.

Leo Computers Ltd. was absorbed in 1963 into English Electric Ltd., a company that had been associated with the National Physical Laboratory in Project ACE and had produced and marketed its own computer, DEUCE. Like Lyons, English Electric had become interested in computers because of "the company's own need for in-house digital computing facilities." After manufacturing a number of commercial machines, English Electric merged with another British computer company (ICT); the new company was known as ICL, for International Computers Ltd (92).

The LEO computers were used by many different types of businesses and organizations. They produced life tables for annuities and range tables for artillery for the armed forces. They solved problems for the aircraft industry and the Coal Board. One problem, involving all the railroad connections in Britain (some 4,000 stations), would have taken fifty clerks five years of calculation by hand.

There was also in Britain, as in America, a considerable input from the government. The British Ferranti Mark I and its successors (1951–1954) were produced with "the impetus of government money and the access to University research." I do not know how the Manchester "Mark I" — also known as MADM (Manchester Automatic Digital Machine) and MUC (Manchester University Computer) — was financed (93). The Pilot ACE was a government project at the National Physical Laboratories (94). It would be valuable to know more about governmental (and especially military) involvement in other British computer developments, by ICT and ICL, Elliott Brothers, and Ferranti, and the computers at Manchester (95).

Notes

1. These remarks apply especially to developments in the United States. Some information on the rather different situation in Europe is given below.
2. In dealing with the early history of computers, there is a vexing problem of terminology. Prior to World War II, and during the war years, a "computer" was a person sitting at a desk and using a desk calculator. Many historians and computer scientists today reserve the term "computer" for a machine that is both digital and all-electronic and that embodies the "stored program." This definition leads to confusions and problems, however, since ENIAC was officially designated by its designers as a "computer," although it had no stored program; contrariwise, the first digital, electronic machine to be put into regular full-time operation, embodying the stored program, was EDSAC, at Cambridge University, which was officially called a "calculator." An additional complication arises from the fact that there are some machines currently being made that are called "computers," but which embody the concept of "Harvard architecture" and do not make use of the stored program. Accordingly, I have not used the restricted definition of "computer" here, but have referred to the early machines of the 1940s indiscriminately as "calculators," "supercalculators," or "computers."
3. Aiken bitterly recalled the attitude of the Harvard Physics Department and of President Conant during the course of an oral-history interview that Henry Tropp

and I conducted with him in his home at Fort Lauderdale, Florida, 24 February 1973.

4. The attitude toward ENIAC, and the Moore School generally, on the part of "established scientists" and those who later came to be called "gate keepers," is described in full and analyzed on pp. 16—23 of Nancy Stern, *From ENIAC to UNIVAC: An Appraisal of the Eckert-Mauchly Computers* (Bedford, Mass.: Digital Press, 1981).
5. The exploration in full of the machines used for encription, decription, and code-breaking in the United States during World War II would be a most valuable research project.
6. See D.J. Frailey, "Computer Architecture," in Anthony Ralston (ed.), *Encyclopedia of Computer Science and Engineering*, 2nd ed. (New York: Van Nostrand Reinhold Company, 1983), pp. 275—288.
7. Stern, *op. cit.*, 1981 (4), Chap. 6.
8. See the work on Aiken cited in n. 10, infra.
9. A major listing may be found in Saul Rosen, "Electronic Computers: A Historical Survey," *Computing Surveys* **1** (1969), 7—36. See also U.S. Navy, Office of Naval Research, *Digital Computer Newsletter*, nos. 1—3, 1951; also U.S. Navy, Office of Naval Research, *A Survey of Automatic Digital Computers* (1953). Some characteristics of early machines are listed in M. V. Wilkes, "Digital Computers: History (Early)," in Ralston, *op. cit.*, 1983 (6), pp. 535—540.
10. Maurice V. Wilkes, talk given at the Pioneer Day celebration commemorating the contributions of Howard H. Aiken, National Computer Conference (AFIPS), Anaheim, Calif., 1983: the substance of this talk will be incorporated in a chapter in I. B. Cohen (ed.), *Howard H. Aiken, Computer Pioneer* (to be published by the MIT Press). See, further, M. V. Wilkes, *Memoirs of a Computer Pioneer* (Cambridge, Mass.: The MIT Press, 1985).

    While radar was the chief producer of experience in electronics, there were other fields which also provided technical skills in devising and using electronic circuits, notably code-breaking and fire control.
11. Herman H. Goldstine, *The Computer from Pascal to von Neumann* (Princeton: Princeton University Press, 1972), p. 251. Funding for the IAS came from the Office of Naval Research, Army Ordnance, and a "tri-service contract which for a while was supplemented by the Atomic Energy Commission."
12. *Proceedings of a Symposium on Large-Scale Digital Calculating Machinery*, Jointly Sponsored by the Navy Department Bureau of Ordnance and Harvard University, at the Computation Laboratory, 7—10 January 1947, vol. 16 of *Annals of the Computation Laboratory of Harvard University* (Cambridge, Mass.: Harvard University Press, 1948); reprint ed. with a new introduction by William Aspray, vol. 7 of Charles Babbage Institute, Reprint Series in the History of Computing (Cambridge, Mass.: The MIT Press; Los Angeles: Tomash Publishers, 1985). *Proceedings of a Second Symposium on Large-Scale Digital Calculating Machinery*, Jointly Sponsored by the Navy Department Bureau of Ordnance and Harvard University at the Computation Laboratory, 13—16 September 1949, vol. 26 of *Annals of the Computation Laboratory of Harvard University* (Cambridge, Mass.: Harvard University Press, 1951).
13. Among Americans only J. Presper Eckert and John von Neumann were missing from the list, which included many other Americans and also Europeans; but

Eckert and von Neumann were represented by associates. At the last minute, Norbert Wiener withdrew.
14. Martin Campbell-Kelly and Michael R. Williams (eds.), *The Moore School Lectures: Theory and Techniques for the Design of Electronic Digital Computers*, vol. 9 of Charles Babbage Institute, Reprint Series in the History of Computing (Cambridge, Mass.: The MIT Press; Los Angeles: Tomash Publishers, 1985).
15. University Mathematical Laboratory, Cambridge, *Report of a Conference on High Speed Automatic Calculating-Machines, 22—25 June 1949* (Cambridge: Issued by the Laboratory with the Co-operation of the Ministry of Supply, January 1950; reprinted 1977).
16. Simon Lavington, *Early British Computers* (Bedford, Mass.: Digital Press, 1980), pp. 23—26.
17. *Ibid.*, chaps. 5, 10.
18. *Ibid.*, p. 52.
19. Uta Merzbach, *George Scheutz and the First Printing Calculator*, Smithsonian Studies in History and Technology, no. 36 (Washington, D.C.: Smithsonian Institution Press, 1977).
20. Anthony Hyman, *Charles Babbage, Pioneer of the Computer* (Princeton: Princeton University Press, 1982), pp. 52—53, 123—128.
21. Larry Owens, "Vannevar Bush and the Differential Analyzer: The Text and Context of an Early Computer," *Technology and Culture* **27** (1986), 63—95.
22. D. H. Sadler, "Comrie, Leslie John," *Dictionary of Scientific Biography*, III, 373a—374b.
23. Michael R. Williams, *A History of Computing Technology* (Englewood Cliffs, N.J.: Prentice-Hall, 1985), pp. 181—182.
24. Geoffrey Austrian, *Herman Hollerith: Forgotten Giant of Information Processing* (New York: Columbia University Press, 1982).
25. See Charles J. Bashe, Lyle R. Johnson, John H. Palmer and Emerson W. Pugh, *IBM's Early Computers* (Cambridge, Mass.: The MIT Press, 1986), chap. 1.
26. Jean Ford Brennan, *The IBM Watson Laboratory at Columbia University: A History* (n.p.: International Business Machines Corporation, n.d.).
27. For an excellent, concise presentation of computation with punched cards, see Bashe et al., *op. cit.*, 1986 (25) chap. 1.
28. A good survey of the classic analog computers may be found in Ellice Martin Horsburgh (ed.), *Modern Instruments and Methods of Calculation: A Handbook of the Napier Tercentenary Celebration Exhibition* (London: G. Bell and Sons, 1914); reprint ed., vol. 3 of Charles Babbage Institute, Reprint Series in the History of Computing (Cambridge, Mass.: The MIT Press; Los Angeles: Tomash Pulishers, 1982). See also Williams, *op. cit.*, 1985 (23), chap. 5, "The Analog Animals."
29. See Williams, *op. cit.*, 1985 (23), pp. 206 —212.
30. George R. Stibitz, "Computer," first published from a manuscript of 1940 in Brian Randell (ed.), *The Origins of Digital Computers: Selected Papers*, 2nd ed. (New York: Springer-Verlag, 1975), pp. 241—246.
31. Williams, *op. cit.*, 1985 (23), §6.3.
32. Stibitz, *op. cit.*, 1975 (30).
33. Williams, *op. cit.*, 1985 (23), p. 228, where a full description of the system may be found.

34. The successors to the Bell Model I (that is, Models II—VI) are described briefly below. The best account in print of Stibitz's machines is to be found in Williams, *op. cit.*, 1985 (23) §6.3.
35. George R. Stibitz, "Introduction to the Course on Electronic Digital Computers" [1946], in Campbell-Kelly and Williams, *op. cit.*, 1985 (14), p. 9. Stibitz also described how "most mathematicians were willing to leave their ivory towers and come down to earth. In other words, they were willing, if not exactly anxious, to soil their hands with computation. Time was, when the supreme insult to a mathematician was to suggest that he find a numerical solution to a problem. Now, however, he was not opposed to casting his work in a form which would lead to a calculated, numerical answer that could actually be used. Many of us came to the conclusion that it was just as noble (during the war, that is) to express it in terms of a convergent but incalculable series of tabulated functions, or to prove that such a series did not exist" (*Ibid.*).
36. See Williams, *op. cit.*, 1985 (23), pp. 231 —232; also Stan Augarten, *Bit by Bit: An Illustrated History of Computers* (New York: Ticknor & Fields, 1984), pp. 99—102. This machine was later turned over to the U.S. Naval Research Laboratory, where it remained in operation until 1961. For a description, see O. Cesareo, "The Relay Interpolator," *Bell Laboratories Record* **23** (1946), 457—460; reprinted in Randell, *op. cit.*, 1975 (30), pp. 247—250.

    See, further, George R. Stibitz, "Early Computers," in N. Metropolis, J. Howlett, and Gian-Carlo Rota, *A History of Computing in the Twentieth Century* (New York: Academic Press, 1980), pp. 479—483; *idem*, "The Relay Computers at Bell Labs" (as told to Evelyn Loveday), *Datamation*, **13**, nos. 4, 5 (1967), 35—44, 45—49; *idem*, "Computer" (unpublished memorandum, 1940), in Randell, *op. cit.*, 1975 (30), pp. 241—246; I have also had the advantage of reading a draft of Stibitz's MS autobiography, tentatively entitled *The Zeroth Generation*, which — it is hoped — may eventually be published by the MIT Press. See also G. R. Stibitz and J. A. Larrivee, *Mathematics and Computers* (New York: McGraw-Hill, 1957).
37. Completed in June 1944, this machine was moved to the Army Field Forces Board in Fort Bliss, Texas, in 1948, where it remained in operation until 1958. It was succeeded by another Ballistic Computer, called the Bell Laboratories Relay Calculator Model IV (known in the Navy as Error Detector Mark 22); completed in 1945, it remained in the service of the Navy until 1961. Bell also constructed a further development in the Model V (a "twin machine"), which was quite properly described as an "all-purpose computing machine"; see F. L. Alt, "A Bell Telephone Laboratories' Computing Machine," *Mathematical Tables and Other Aids to Computation* **3** (1948), 1—13, 69—84, reprinted in Randell, *op. cit.*, 1975 (30), pp. 257—286.

    After the war, Bell built the last of the series, a Model VI, for its own use at its Murray Hill center. In the late 1950s the Model VI was given to the Polytechnic Institute of Brooklyn. The latter presented the machine in 1961 to the Bihar Institute of Technology in India, "where it still resides [1985] as an historical display"; see Williams, *op. cit.*, 1985 (23), pp. 231—240. For details of the Model VI, see E. G. Andrews, "The Bell Computer, Model VI," *Proceedings of a Second Symposium* (see n. 12 supra), pp. 20—31; according to Andrews, "Previous Bell computers have not carried model numbers"; the "numbers that are now assigned to them" have been given "in retrospect." On the Model III, see Joseph Juley, "The Ballistic Computer," *Bell Laboratories Record* **23** (1947), 5—9, reprinted in Randell, *op. cit.*, 1975 (30), pp. 251—255.

38. See John V. Atanasoff: "Computing Machine for the Solution of Large Systems of Linear Algebraic Equations" (unpublished memorandum, 1940), published in Randell, *op. cit.*, 1975 (30) pp. 305—325.
39. In a "financial statement" (*ibid.*, pp. 324— 325), Atanasoff accounted for receipts of $1,460 (including a grant from "Dean Gaskill, Winter, 1940" of $110); he eventually hoped "to secure a grant of $5,000 in addition to the balance already allotted to the project, to be used during the next two years to complete the construction of the machine, and to test and perfect it under actual operating conditions."
40. See Stern, *op. cit.*, 1981 (4), pp. 3—5.
41. The most recent contribution to this controversy has just been announced for publication by the University of Michigan Press: Alice R. Burks and Arthur W. Burks, *The First Electronic Computer: The Atanasoff Story* (advertised in the book review section of the *New York Times* for publication in January 1988).
42. Aiken's original prospectus, "Proposed Automatic Calculating Machine," dated 4 November 1937, is available in Randell, *op. cit.*, 1975 (30), pp. 191—197.
43. Before going to IBM, Aiken went to the Monroe Calculating Machine Company, where the chief engineer, George Chase, was very anxious to build a machine according to Aiken's general plan. Chase did not succeed in convincing his colleagues that they should build the machine and recommended that Aiken try IBM. When I discussed this topic with Aiken, during the oral-history interview that Henry Tropp and I conducted on 24 February 1973, Aiken said that he knew from the start that his machine was going to cost "a lot of money" to build. "Thousands and thousands of parts! It was very clear that this thing could be done with electronic parts, too, using the techniques of the digital counters that had been made with vacuum tubes, just a few years before I started, for counting cosmic rays. But what it comes down to is this: if Monroe had decided to pay the bill, this thing would have been made out of mechanical parts. If RCA had been interested, it might have been electronic. And it was made out of tabulating machine parts because IBM was willing to pay the bill." Quoted by I. B. Cohen, *Annals of the History of Computing* **2** (1980), p. 200.
44. It is thus incorrect to say (see Williams, *op. cit.*, 1985 [23], p. 241) that Aiken "managed to convince IBM that, with financial help from the United States Navy, it should undertake to construct a large-scale computing machine." For an "official" presentation of IBM's participation in this project see the work by Bashe et al., *op. cit.*, 1986 (25).
45. See Aiken's prospectus (42), p. 192.
46. The actual problems assigned to Mark I during the war years are described by Robert V. Campbell and Richard Bloch, programmers of Mark I in that period, in Cohen, *op. cit.*, (10).
47. Mark I ran almost continuously during the war years (and the years immediately following): 24 hours a day, 7 days a week. Any free time not allocated to specific problems, was used for what Aiken called "computin'" or "makin' numbers" — the production of tables, which were subsequently published in the *Annals of the Computation Laboratory*.

48. The best accounts of Zuse's work in English are in Paul Ceruzzi, *The Reckoners* (Westport, Conn.: Greenwood Press, 1983); Augarten. *op. cit.*, 1984 (36), pp. 88—97; and Williams, *op. cit.*, 1985 (23), §6.2. Zuse has several publications in English: "Method for Automatic Execution of Calculations with the Aid of Computers" (1936), in Randell, *op. cit.*, 1975 (30), pp. 159—166; "The Outline of a Computer Development from Mechanics to Electronics" (1962), in *ibid.*, pp. 171—186; and "Some Remarks on the History of Computing in Germany," in Metropolis, Howlett, and Rota, *op. cit.*, 1980 (36), pp. 611—627. Zuse has also written an autobiography, *Der Computer — Mein Lebenswerk* (Munich: Verlag Moderne Industrie, 1970).
49. Randell, *op. cit.*, 1975 (30), p. 156.
50. The Z3 used 2,600 electromagnetic relays. Its memory was very limited, with a maximum capacity of sixty-four 22-bit numbers. It could perform the four basic operations of arithmetic, extract square roots, and do other numerical tasks, but — like Aiken's Mark I — it was unable to make conditional jumps. There was no provision for branching circuits. Williams points out that despite its many extraordinary features, the Z3 was limited to a "modest four-decimal-digit accuracy," whereas Aiken's Mark I "could deal with 23 digits"; and the Z3 had "no internally stored program" (*op. cit.*, 1985 [23], p. 221).
51. The flying bombs produced by Henschel were not the V-series of rockets, carrying large bombs, with which Germany attempted to destroy Great Britain during the last period of the war. Rather, these flying bombs were transported by airplanes and released near their targets, being guided by radio from the aircraft. See Augarten, *op. cit.*, 1984 (36), p. 96.
52. See Williams, *op. cit.*, 1985 (23), pp. 222 —224.
53. *Description of a Relay Calculator*, by the Staff of the Computation Laboratory, vol. 24 of *Annals of the Computation Laboratory of Harvard University* (Cambridge, Mass.: Harvard University Press, 1949).
54. See Brennan, *op. cit.*, (26).
55. Williams, *op. cit.*, 1985 (23), pp. 225—226. This machine continued in use, with various modifications (models 602, 602A, 603, 604, and 605) for many years. The 604 was the first of this series in which vacuum tubes replaced the relays. See, further, Bashe et al., *op. cit.*, 1986 (25).
56. Williams, *op. cit.*, 1985 (23), p. 259. On this subject, see also Bashe et al., *op. cit.*, 1986 (25), chap. 2, p. 45.
57. Williams, *op. cit.*, 1985 (23), pp. 259—260; Bashe et al., *op. cit* 1986 (25).
58. Williams, *op. cit.*, 1985 (23), p. 260. See, further, W. J. Eckert, "The IBM Pluggable Sequence Relay Calculators," *Mathematical Tables and Other Aids to Computation* **3** (1948), 149—161. An informative source on the various computing projects during World War II is Bernard O. Williams, *Computing with Electricity* (unpublished doctoral dissertation, University of Kansas, 1984).
59. Williams, *op. cit.*, 1985 (23), p. 272.
60. See Owens, *op. cit.*, 1986 (21).
61. See Goldstine, *op. cit.*, 1972 (11), pp. 130—139.
62. The most complete history of ENIAC from conception to operation is Stern, *op. cit.*, 1981 (4). Mauchly's document is currently in the University of Pennsylvania Archives; its significance is discussed by Stern in footnote 20 on p. 256.

63. Williams, *op. cit.*, 1985 (23), p. 287.
64. On von Neumann's association with ENIAC, see Stern, *op. cit.*, 1981 (4), and Goldstine, *op. cit.*, 1972 (11). Neither of these authors, however, mentions the fact that even before von Neumann's famous encounter with Goldstine and his subsequent introduction to ENIAC, he had already spent time at Harvard, where Richard Bloch had programmed for him (on Mark I) an implosion problem for the atom bomb.
65. For this document, see Randell, *op. cit.*, 1975 (30), pp. 355—364. The first complete printing of von Neumann's "First Draft of a Report on the EDVAC" is to be found in chapter one of William Aspray and Arthur Burks (eds.), *Papers of John von Neumann on Computing and Computer Theory*, vol. 12 of Charles Babbage Institute, Reprint Series for the History of Computing (Cambridge, Mass.: The MIT Press; Los Angeles: Tomash Publishers, 1987).
66. Randell, *op. cit.*, 1975 (30), p. 350.
67. On the Williams tube, see *ibid.*, pp. 387—388.
68. *Ibid.*, p. 352.
69. On the Moore School lectures, see Campbell-Kelly and Williams, *op. cit.*, 1985 (14). EDSAC will be discussed in the final section, below.
70. Brian Randell's preliminary report on his explorations of the secret history of Colossus is a lecture given at the Los Alamos Computer History Conference, 10—15 June, 1976; this was published in Metropolis, Howlett, and Rota, *op. cit.*, 1980 (36), pp. 47—92.
71. The full report on Colossus was, at last, published in *Annals of the History of Computing* **5**, no. 3 (1983), including: Thomas H. Flowers, "The Design of Colossus," pp. 239—252; Allen W. M. Coombs, "The Making of Colossus," pp. 253—259; and W. W. Chandler, "The Installation and Maintenance of Colossus," pp. 260—262.
72. Flowers, *op. cit.*, 1983 (71), p. 252.
73. On the IBM 701, see the special issue of *Annals of the History of Computing* (**5**, no. 2 [April 1983]), edited by Cuthbert C. Hurd, devoted to this machine. At the time when IBM completed and marketed the 701, the priority system for raw materials meant that only potential users who had missions tied to the war effort could obtain the new calculator. Nineteen 701s were constructed; the first was installed in IBM's World Headquarters. The pattern of distribution of the others shows a predominance of use of machines (ten 701s) in the aircraft and allied industries; there were also three in weapons laboratories, and one each in the National Security Agency, the Rand Corporation — associated with the Air Force — and a Navy installation. Production of the 701 actually ceased with no. 18, but eight months later a nineteenth was specially assembled from spare parts for use in meteorological predictions — in a project sponsored jointly by the Weather Bureau, the Air Force, and the Navy at the U.S. Weather Bureau at Suitland, Maryland.
74. IBM offered to undertake the development and construction of NORC for the Naval Ordnance Laboratory "on a research and development basis, under which we would be reimbursed for our costs plus a fixed fee" (the fee was later set at $1). IBM's official technical history records that the "principal incentive for IBM was

experience in design, construction, and maintenance" of "a machine that would employ the latest and most promising technologies"; but "other incentives were potential patents, customer goodwill, and favorable publicity." The Navy funds for support of NORC came to approximately $2.5 million. A later assessment by IBM of design and construction features of NORC showed that the best way to profit by the new techniques was not to use NORC hardware or system components in commercial systems, but rather to make "a strong transfusion" of "experienced NORC personnel into product-oriented groups." See Bashe et al., *op. cit.*, 1986 (25), pp. 133, 181—183.

75. In February 1956 IBM proposed to design and construct STRETCH for the Los Alamos Scientific Laboratory for $4.3 million. To gain some idea of the magnitude of such a project, and to see how it could not be undertaken within IBM's regular resources, we may turn to an *IBM Poughkeepsie Laboratory Newsletter* (24 April 1956) that states: "We have sent 200 lecturers directly to some 37 campuses." The technology developed for STRETCH saw almost immediate service in IBM's product line in the 7090 and its successor 7030 systems and in a super-STRETCH or HARVEST machine for NSA. See Bashe et al., *op. cit.*, 1986 (25), chap. 11.
76. See Stern, *op. cit.*, 1981 (4).
77. A history of the involvement of ARPA (Advanced Research Projects Agency, now DARPA) has been commissioned from the Charles Babbage Institute (Arthur Norberg, Director; William Aspray, Associate Director); this investigation will be limited to DARPA's Information Processing Technology Office (IPTO).
78. Parts of the following discussion of ONR in relation to the development of computers draw heavily on, and summarize, an unpublished research report by Elizabeth C. Luebbert: "The Development of the Digital Computer: A Case of ONR Support of Research" (6 May 1974, Harvard University, John F. Kennedy School of Government: Seminar in Science, Technology, and Public Policy). This has been supplemented by two published reports by Mina Rees: "The Computing Program of the Office of Naval Research, 1946—1953," *Annals of the History of Computing* **4** (1982), 102—120; and "The Mathematical Sciences and World War II," *American Mathematical Monthly* **87** (1980), 607—621. A full-scale history of ONR would be a most welcome and important addition to knowledge.
79. See A. Hunter Dupree, *Science in the Federal Government* (Cambridge: Harvard University Press, 1957), p. 375.
80. ONR was formally established on 1 August 1946 (Public Law 588) by the 79th Congress, 2nd Session. But even earlier, in September 1945, Admiral Forrestal had created by executive order an Office of Research and Inventions, using funds from the regular budget. This organization was dissolved when ONR came into being. It was instituted because of the Navy's concern for the health and well-being of research science, especially the needs for basic science.
81. Waterman's academically oriented programs and attitude were continued by Emanuel R. Piore, who succeeded Waterman at ONR and later became Director of Research and Chief Scientist (and a vice-president) at IBM.
82. Luebbert, *op. cit.* (78).
83. *Ibid.* See also Kent C. Redmond and Thomas M. Smith, *Project Whirlwind: The History of a Pioneer Computer* (Bedford, Mass.: Digital Press, 1980).

84. Luebbert, *op. cit.* (78).
85. Goldstine, *op. cit.*, 1972 (11), p. 331.
86. I am, once again, referring primarily to the American experience; for a somewhat different pattern of development in Britain, see below.
87. Quoted in Goldstine, *op. cit.*, 1972 (11), p. 331.
88. A real lacuna in the history of computers is information on the architecture, functioning and programming, financial costs and support, and personnel of European machines. See William Aspray, "International Diffusion of Computer Technology," *Annals of the History of Computing* **8** (1986), 351—360.
89. The EDSAC (Cambridge, England, 1949), the first digital all-electronic stored-program to be put into permanent operation, was built with a small budget. Financial support in the form of staff salaries came from University funds, presumably derived ultimately from the University Grants Commission, plus a small grant from Lyons.

    In his *Memoirs, op. cit.*, 1985 (10), Maurice Wilkes describes the construction of the EDSAC but does not indicate the source of external funds, save for a grant from Lyons of £3000 and "the services of an assistant for a year." Simon Lavington, says of EDSAC that the "financing of the project was via normal University research channels, plus a (for those days) sizeable donation of £2500 from D. Lyons & Co. Ltd," *op. cit.*, 1980 (16).
90. Lavington, *op. cit.*, 1980 (16), chap. 13, quotes extensibly from Thompson's unpublished "LEO Chronicle." See also *ibid.*, p. 35.
91. The information in Lavington has been supplemented by a lecture given by Pinkerton at the Computer Museum, Boston, Mass., on 5 October 1987, and by discussion with Pinkerton after the lecture. This lecture has been published in *The Computer Museum Report* **21** (1987—1988), 11—16; its title is "The Early History of LEO: The First Data Processing Computer."
92. Lavington, *op. cit.*, 1980 (16), p. 35 and chap. 14.
93. An excellent survey of "The British Scene" is given in chap. 8 of Williams, *op. cit.*, 1985 (23).
94. Lavington, *op. cit.*, 1980 (16), appendix 4, gives "technical specifications of 15 early British computers." See also chap. 9 ("Defence Computers") and the chapters on Elliott Brothers, Leo, and English Electric Computers. A major British military stored-program computer (1947—1954) was MOSAIC (Ministry of Supply Automatic Integrator and Computer).
95. On the Ferranti machines, see Lavington, *op. cit.*, 1980 (16).

STYLES OF MILITARY TECHNICAL DEVELOPMENT: SOVIET AND U.S. JET FIGHTERS – 1945–1960

LEON TRILLING

Massachusetts Institute of Technology

This paper describes and compares the development and the performance of the fighter aircraft built in the Soviet Union and the United States immediately after the Second World War. Their design provided the aeronautics community with much experience, and some of the technical background needed for the postwar expansion of jet aviation; it took place under rather different institutional, incentive, and resource contexts. I try to show how those constraints affected the process of design and the resulting aircraft, which in the end achieved comparable effectiveness by rather different means.

In particular, the discussion focuses on disparities in production facilities and trained manpower; on differences in the punishment-reward system; and on the contrast in the organization of the government and other entities that were responsible for the design, production, and deployment of fighter aircraft.

The Context of Jet Fighter Development

By the late 1930s, L. Caproni in Italy (1) and V. F. Bolkhovitinov in Soviet Union (2) had developed and tested prototypes of single-seat jet aircraft. In 1944–45, jet-propelled operational fighters such as the British Gloster Meteor (3) and the German Messerschmitt Me-262 (4) achieved standards of performance that made propeller-driven aircraft obsolete almost at once in first-line military use, and eventually in commercial air transportation. Political circumstances at the end of World War II also provided strong incentives for both American and

E. Mendelsohn, M. R. Smith and P. Weingart (eds.), Science, Technology and the Military, Volume XII, 1988, 155–185.

Soviet military planners to exploit the potential of jet engines to improve the effectiveness of their respective air forces.

M. Sherwin shows how the expansion of Soviet power and influence in Eastern Europe and the demonstration of the effectiveness of U.S. atomic weapons became linked in the formulation of American security policy in 1945—48 (5). At first, there was the hope that a promise to share some information on the production of atomic energy and to cooperate in its nonmilitary use would provide a possible lever with which to control Soviet expansion into Eastern Europe — particularly into Poland, which was not directly accessible to U.S. power. The failure of the United States to prevent Soviet expansion and consolidation (e.g., the Soviet move into Czecholovakia [1948], and their suppression of nationalistic outbreaks in Poland [1953] and Hungary [1956]), and the ideological interpretation put by most U.S. and British leaders on Soviet policy (e.g., Winston Churchill's "iron curtain" speech at Fulton, Missouri; [1946]), helped to create both a climate of increasing U.S. and British antagonism toward the USSR, and the will to maintain a strong U.S. military posture.

That cold-war mood limited, and eventually reversed, the usual American policy of demobilization and military contract cancellations that had followed the Civil War and World War I and briefly held sway in 1945—47 (6). Yet, the United States was unwilling to remain fully mobilized. The sense of U.S. superiority in most technical fields, based on the complete disarmament of Germany and Japan after the Allied victory of 1945, the economic weakness of the U.K. and the USSR, and the spectacular accomplishment of the atomic bomb, all contributed to support a policy that relied on technical military superiority to offset the numerical superiority expected of Soviet (or Chinese) adversaries.

The Soviets had paid dearly for their 1945 victory over Germany, and their leaders felt themselves to be in a basically unfriendly international environment; the achievement of security and the acquisition of the material resources needed to rebuild the USSR were their highest priorities. The collapse of political authority in Central Europe also seemed to provide some opportunities for expansion. The Soviets pursued both objectives by creating a buffer zone of small states on their western frontier, which they made into reliable allies, in spite of often strong anti-Russian nationalism, by maintaining in power a leadership ideologically sympathetic to the USSR and politically and militarily dependent on it.

At the same time, the Soviet leadership insisted on a strong defensive military posture. This required a large standing army and a strong antiaircraft defense, including very substantial investments in antiaircraft guns in 1945—55 and in ground-to-air missiles in the late 1950s (7). As a part of this defensive posture, the Soviets also maintained a substantial force of simple short-range interceptor fighters to keep a potential enemy (atomic) bomber force from penetrating the buffer that protected the western approaches to the Soviet Union.

Thus, both the United States and the Soviet leaders saw a need for substantial modern air forces, and this required the further development of jet aircraft, which had proved their superior performance at the end of World War II.

The new generation of fighter aircraft might operate at top speeds of well over 600 miles per hour, speeds at which the effects of air compressibility on the performance, the handling characteristics, and the structural integrity of the aircraft are important. All the forces and moments that act on an aircraft increase roughly as the square of the airspeed, and additional (then known) large effects (drag increase, pitching moment shifts, buffeting) occur when the flight speed is comparable to the speed of sound, some 750 miles per hour. (The ratio of flight speed to the speed of sound, called the Mach Number [M], is used by engineers as an index of the importance of compressibility and transonic effects. Below M 0.5 these are negligible; the region between 0.7 and 1.3 is the tricky transonic region; above 1.6, one is in the better-understood supersonic zone. The maximum Mach number of fighters in the period considered here increased from 0.6 to 2.5.) A sense of how the resulting design difficulties were perceived in the 1940s and 1950s, and how urgent it seemed to overcome them, is conveyed in popular technical books such as *The Sound Barrier* by Duke and Lanchberry (8) and, more dramatically, by several motion pictures made by Neville Shute (the pseudonym of J. Norway, a British aeronautical engineer turned movie director), in particular *Breaking the Sound Barrier*, released in 1954.

The graphs in Figures 1 and 2 give some idea of the change in the performance of first-line fighters during the period 1942—60 (9). That

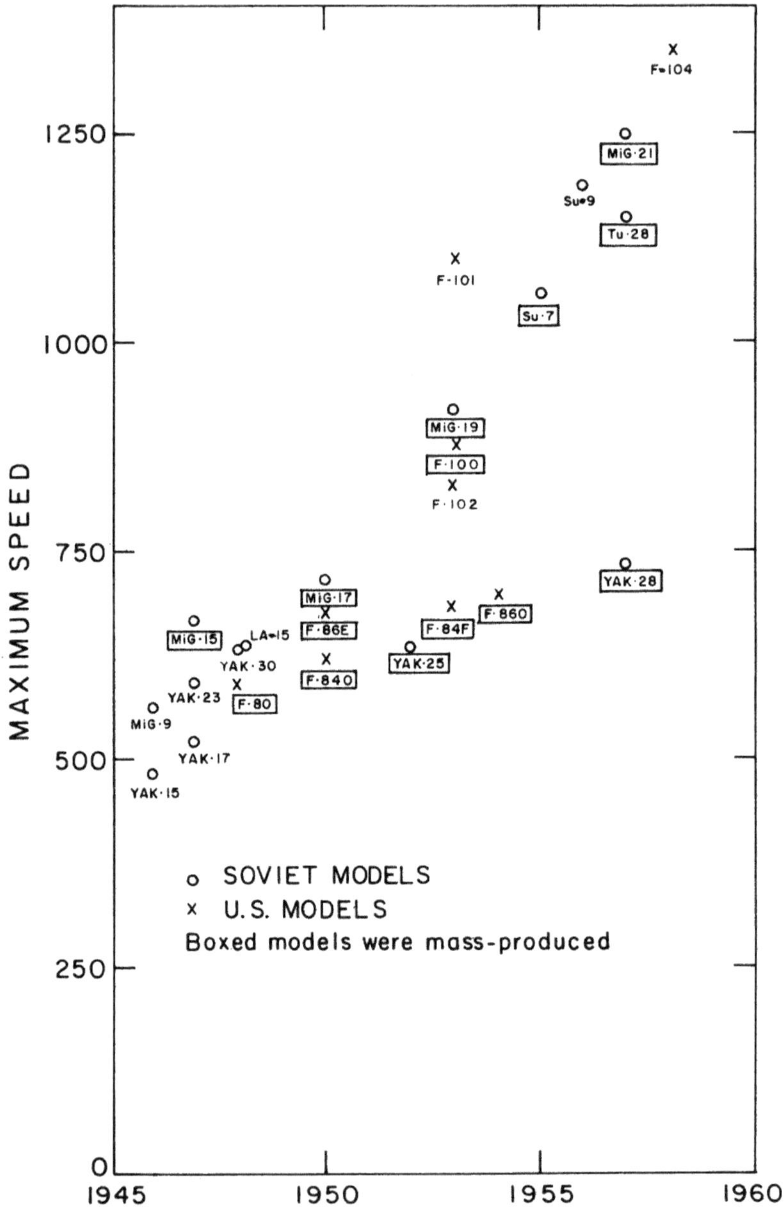

Fig. 1. Maximum speed of some early jet fighters (Source: Ref. 2).

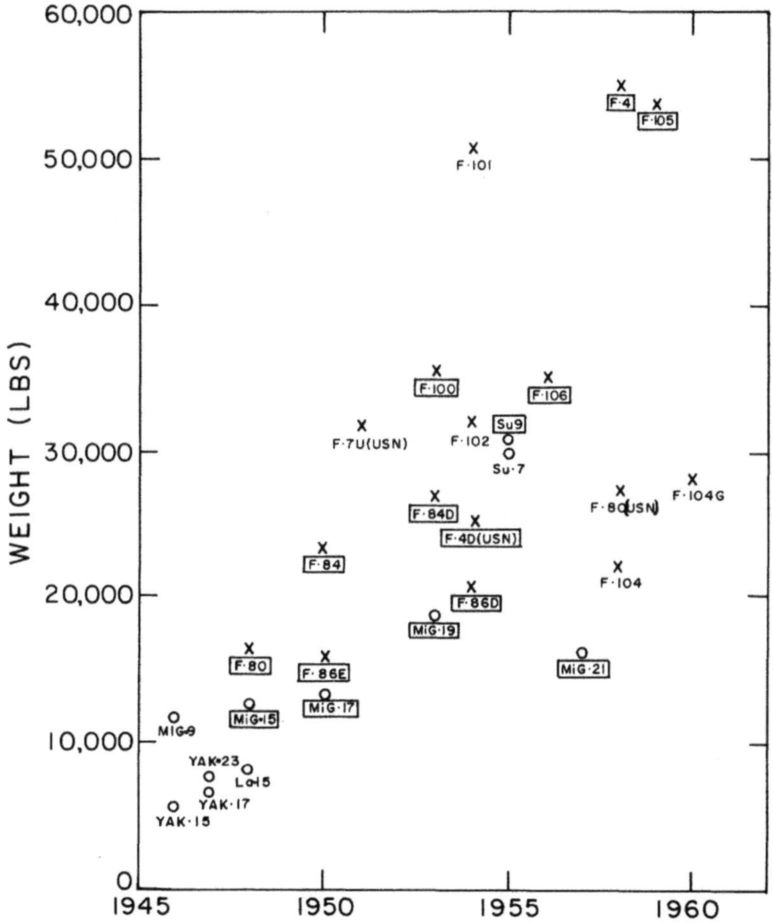

Fig. 2. Weight of typical fighters (Source: Ref. 2).

evolution occurred in three phases. First, between 1942 and 1946, interim designs outfitted traditional fighter frames with early versions of jet engines. For example, the Bell P-59 in the United States (1942) and the MiG-9 and Yak-15 in the USSR (1945—46) looked like propeller-driven aircraft without propellers. These ungainly machines served to acquire practical experience with the new jet technology under the new flight conditions. Then, between 1944 and 1948, the first true jet fighters were designed around reliable jet engines and with compressibility effects in mind: the F-80, F-84, and F-86 in its various versions in the United States, the MiG-15, MiG-17, and MiG-19 in the Soviet Union. These aircraft operated just below Mach 1 and were tested in battle against each other in the Korean War (1950—53). The third stage occurred between 1952 and 1957 and witnessed the design of the first supersonic fighters: the Century Series, of which the F-100 was the most effective in the Unites States, and the MiG-21, Su-7 and Su-9 in the USSR. These aircraft coped with transonic aerodynamic difficulties and operated effectively at speeds of Mach 1.5.

Interestingly, the U.S. Department of Defense, the Soviet Ministry of Aircraft Production, and the American and Soviet air forces went about developing these craft in rather different ways. Technical and manpower constraints, institutional and political context, incentive to innovate, and strategic doctrine all differed, resulting in a set of rather different aircraft of comparable performance.

The Soviet Context and Fighter Development Program

To understand the background in which the Soviet postwar fighters were designed and produced, one must remember the enduring geography of the Soviet Union: the very large spaces connected by a sparse network of roads and railways, the punishing cold and snow of the Russian winter, and the arid desert heat of Central Asian summers. Soviet fighters were designed to operate under extreme weather conditions from primitive fields and runways. The personnel available to maintain and operate these aircraft in the late 1940s were of unpredictable competence because the training program for technicians, begun in the early thirties, had been disrupted by the war and had not yet graduated anywhere near enough mechanics (10).

The country had suffered heavy losses in personnel and severe disruptions in its industrial plant as a result of World War II. Much of the aircraft industry, including its research establishment, had been evacuated beyond the Urals and was returning to its original locations in the slowly rebuilding industrial centers of European Russia; was seriously understaffed and underequipped (11). Indeed, the tooling available to the Soviets as late as 1960 consisted mostly of European and American imports from the 1930s and 1940s, and of locally made copies. By the late 1950s a drive to design and produce automated tooling was under way, and demonstration specimens were being shown to foreign visitors (12), but these were not yet in wide use and could not be counted on to manufacture new designs (13).

Finally, the Soviet Union was coming through an extraordinarily difficult, tense fifteen years. Agriculture had been collectivized at the cost of a severe famine (1930—33), and the first two 5-year plans (1930—40) promoted the development of heavy industry with little regard for consumer needs. A series of purges disrupted the nation's political, military, administrative, intellectual, and technical leadership (1935—38). World War II (1941—45) cost the Soviet Union some ten to twenty million dead and the prolonged foreign occupation of several of the nation's industrial centers. The period from 1946 to 1953 also coincided with Stalin's last paranoid years. This somber background had an important effect on the climate in which military and technical decisions were made.

Up to 1957, Soviet enterprises were organized into fields (e.g., defense, heavy industry, transport, fishing industry, etc.) each administered by a ministry; overall economic goals were defined and at least formally controlled by the State Planning Commission (GOSPLAN) which set yearly and five-year targets and fixed them in practice after complex negotiations with the management of the relevant ministries and plants (14). After 1957, the management of some economic activities was transferred to local agencies, but defense-related production and research remained under central management. In an economy deprived of the market-price mechanism, allocations of resources had to be made by a combination of orders from higher authority and scramble. Therefore, each production ministry tried to gain control of the most critical resources needed for the fulfillment of its functions and of its assignment under "the plan." Officials would supervise the training of their engineers and technicians, deal for the supply of the scarcer raw materials, manufacture critical subassemblies, and closely manage their design and plant facilities. That process created a trend toward vertical integration much like Fordism in America, and it is no accident that Henry Ford was the American entrepreneur most admired by Soviet officials in the 1930s and 1940s.

The defense-related industries had the highest priority in claims for resources and fairly direct access to the top leadership. This was particularly important because intermediate-level bureaucrats were reluctant to make decisions. When a conflict (e.g., over resource allocation) arose, they would refer the decision to higher authority because there was no reward for deciding "right," and the definition of "wrong" remained ambiguous.

The risks of failure were considerable. The aircraft designer D. P. Grigorovich was put under arrest in 1930—33 and disappeared in 1938 (15). The dean of Soviet aircraft designers, A. N. Tupolev, and several of his close associates spent more time in confinement in the late thirties (16), as did S. P. Korolev, the future "chief designer" of Soviet spacecraft, in the forties (17). This did not prevent them from continuing their professional work under secret-police supervision in special camps called "sharagas," well described by A. Solzhenitsyn in *The First Circle* (18).

The Ministry of Aircraft Production (MAP) was one of the ministries that constituted the "Defense Group," headed by a first deputy prime minister. Under its authority several engineering schools — most notably, the Zhukovskii Military Air Academy and the Moscow Aviation Institute — trained most of the aeronautical engineers in the country (19). Actually, after 1957, aviation institutes were under the formal jurisdiction of the Ministry of Higher Education, but their staffs and activities were closely tied to the Ministry of Aircraft Production; the Zhukovskii Academy was under Soviet Air Force jurisdiction. The MAP also supervised several research institutes similar to the Langley and Ames laboratories of the National Advisory Committee for Aeronautics (NACA) and the U.S. Air Force research laboratories at Wright Patterson Air Force Base in the United States. Among the most important of these were the Central Aero-Hydrodynamics Institute (TsAGI), founded in 1924 and, after 1950, located in a suburb of Moscow named Zhukovskii after the scientist N. E. Zhukovskii (1847—1921), who had developed aerodynamic science in Russia; and the Aero-Engine Institute (TsIAM), also located in Moscow (20).

The design of the new aircraft was the task of experimental design teams (Opytnoyie Konstrukorskoyie Buro – OKB) who also built the prototypes at pilot plants under their control, participated with the air forces (or the civil aviation administration) in testing them, and delivered production blueprints to appropriate manufacturing plants. Each OKB usually specialized in the design of a particular type of aircraft, though occasional assignment shifts occurred. The OKBs that had designed most of the Soviet fighters in World War II were headed by S. A. Lavochkin, A. S. Yakovlev, and the team of Artem Mikoyan and Gurevich, better known in the West by the acronym "MiG."

Series construction teams (Seryinoyie Konstrukorskoyie Buro – SKB) were responsible for the large-scale production of approved models according to OKB blueprints. In general, a given SKB and its assembly plant remained in partnership with the same OKB for a long period of time. This helped promote understanding and mutual loyalty between engineering staffs.

A few leaders with multiple appointments coordinated these activities. Some were party officials, military technical officers, and managers, such as Defense Production Ministers K. Rudnev and D. Ustinov and Aircraft Production Minster P. Dementiev; some were senior designers (e.g., aircraft designer A. N. Tupolev, and engine designers V. Klimov and B. Stechkin, who were also members of the Academy of Sciences; designer A. Mikoyan, whose brother was a leading member of the government and a party Politburo; and A. S. Yakovlev, who served for a time as deputy minister); and some were scholars (e.g., Professor M. V. Keldysh, a group leader at TsAGI, who was a member and eventually the president of the Soviet Academy of Sciences). Some of these leaders shared the distinction of being graduates of the Zhukovskii Military Air Academy with the senior officers of the Air Force; several lectured there, as well as at the Moscow Aviation Institute. The personal connections that these "multiple hat" holders provided between the academic and research community (especially the Mechanics Department of Moscow University, the Physico-Technical Institute, the Mechanics Institute, and the Computation Center of the Academy of Sciences), the Air Force, and the top figures of the government were essential for making technical decisions and having them stick (21).

At a conference held at the Kremlin in December 1945 and attended by senior officials including J. V. Stalin, the Soviet leaders decided that a jet fighter of radicallly new design was needed, and all three fighter design teams (MiG, Yak, La) were instructed to take a hand (22). The first set of prototypes designed by Yakovlev and Lavochkin consisted of conventional airframes in which either a German Jumo-002 axial jet engine or an early British centrifugal-flow jet engine was fitted. In the summer of 1947, the Soviets acquired a Rolls Royce Nene engine. This was the most modern, up-to-date engine then available; it developed a thrust of some 5,000 lbs. twice as high as the Jumo engine. Around that engine the MiG team designed a completely new aircraft, of which a prototype was test-flown in December 1947 and a production model was ready in March 1948. In summer 1948, the Soviets purchased twenty-five Nene engines from Rolls Royce along with blueprints and a production license. (At the same time, Rolls Royce also sold a license to the Pratt & Whitney division of United Aircraft Corporation.) V. Klimov's engine design team eventually refined and improved the Nene (1948—50), and the thrust of the resulting VK-1 engine was built up from 5,000 lbs. to 7,500 lbs.

The outcome of these developments was a simple, light, maneuverable, high performance fighter, the MiG-15, which became the standard Soviet air defense fighter (1948—51). A very large number of MiG-15s were built — over 15,000, including several thousand under license in Czechoslovakia and Poland — and the allies of the Soviet Union received several squadrons of them, in particular the People's Republic of China and North Korea, which used them in combat during the Korean War (1950—53) (23).

After the selection of the MiG-15 as the fighter to be mass-produced, Lavochkin and Yakovlev left the field of fighter design. Lavochkin was apparently assigned to the design of ground-to-air missiles; he died in 1960. Yakovlev, after a stint as vice-minister of aircraft production,, designed a series of radar-equipped all-weather interceptors (Yak-25 and its successors); his OKB also produced several helicopters and a light jet transport. The Mikoyan-Gurevich team stayed on as the Soviet Union's leading fighter design group.

The solution of some of the aerodynamic problems encountered in flight at high subsonic speed (550 to 650 mph) had been worked out at the Aero-Hydrodynamics Institute (TsAGI) during World War II, and by 1946 these solutions were available to all designers. For example, the increase in pressure drag, which occurs near the speed of sound, is reduced by sweeping back the wing and horizontal tail surfaces. But the airflow on the upper surface of swept-back wings tends to slide laterally toward the wing tips, reducing the lift and increasing the danger of stall. American designers coped with this problem in several ways: by subtle changes in wing geometry (North American Aviation), by roughening a critical wing surface region (Boeing), or by positioning "fences" parallel to the flight direction to prevent lateral flow. All Soviet fighters of the period used fences; so did their bombers (Il-28, Tu-16), and their jet transports (Tu-104, Tu-134, etc.). This recourse to a single solution worked out at central facilities reflects the acute shortage of large high-performance wind-tunnel facilities in the USSR in the period 1945—60, aggravated by the temporary evacuation of TsAGI from the Moscow area to Siberia during the war, and its gradual return to Zhukovskii after 1945.

Similarly, almost all Soviet aircraft of that period have a T-tail design: the horizontal tail surface is located above the vertical tail fin to keep it out of the jet exhaust. This design simplifies the aerodynamics and accepts a structural penalty; the vertical tail surface must carry loads due to the forces and moments that act on the horizontal surface, and it must be stiffer and heavier to do so.

Sir Arnold Hall, the director of the Royal Aircraft Establishment who visited the USSR in 1956, commented on the "economy of design" and the "satisficing" philosophy (i.e., making devices no more complex than they need to be to serve their intended purposes) of Soviet designers in a tone of approval — possibly explained by the difficulties of the British aircraft design community, which was divided at that time among several fiercely competing small firms who produced a multiplicity of prototypes, none thoroughly tested or put into large-scale production (24).

The MiG team was in sole charge of fighter design after 1948. Some two years after the MiG-15's introduction into service it began to be replaced by the MiG-17, very similar in appearance but with a redesigned wing structure and a larger centrifugal-flow engine (the VK-1 mentioned above) to improve its performance. The design of the MiG-17 was undertaken in 1948; flight-tests took place in 1950, and production began in 1952. Altogether, over 10,000 MiG-17s were built (25). Two years later the MiG-19 appeared: it was a longer, more slender aircraft, shaped for transonic flight and powered by a pair of new axial-flow engines, but it proved to be disppointing in performance. Some 3,000 were built in Czechoslovakia and China. Each of the MiG fighters embodied some moderate improvement over its predecessor and was produced in large numbers to reequip air-force squadrons. Soviet strategic doctrine required a substantial number of modern fighters in actual operation. After their World War II experience the Soviets were very serious about a solid defensive posture, though they may have played a "Potemkin Village" (26) game with long-range bombers and ballistic missiles in the 1950s.

These MiG fighters were produced in large numbers in factories equipped with light tooling of mixed quality. The textbook on aircraft plant management by V. I. Tikhomirov shows that the production lines for which young managers were trained in the late fifties relied on semiautomatic machine tools (lathes, milling machines, etc.) for the fabrication of individual parts, assembly-line production of components, and hand assembly of aircraft by riveting or similar techniques using fixed jigs in World War II style (27). Under such conditions, a certain roughness and large tolerances had to be accepted. These would increase the drag of the aircraft. In order to obtain satisfactory performance a combination of low weight and a large engine was required. The thrust-to-weight ratio of Soviet fighter aircraft of that period is some 50 percent higher than that of U.S. fighters, mostly because the Soviet aircraft were lighter than their American counterparts (see fig. 3). Thus, available production technologies led to a distinct Soviet fighter design style.

The Soviets were capable, we should emphasize, of highly refined, accurate machining of critical parts. The improvement of the British "Nene" engine mentioned above was largely the result of such selective attention to detail (28). But in 1946—55 the Soviets could neither finish nor assemble large component parts with great precision, and

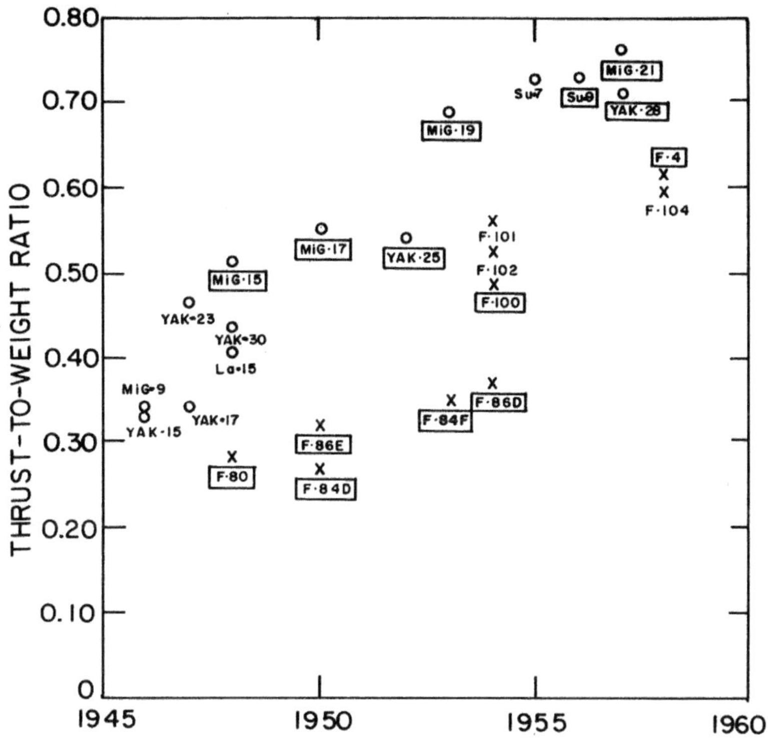

Fig. 3. Thrust-to-weight ratio of typical U.S. and Soviet fighters (Source: Ref. 2).

apparently, in contrast with U.S. Air Force policy, they did not make the development of large, automatically controlled machine tools a high priority until the late 1950s.

The MiG-19 was the last descendant of the MiG-15 line, in which the speed of sound had been reached and slightly exceeded. By 1953, it was time for another step forward — to supersonic performance. This required a heavier structure, greater power, and wings of different geometry: they had to be thinner and to have sharper leading edges. Since the aerodynamic problems of supersonic flight were not yet fully understood, a new competition was advisable in order to hedge risks and to improve performance (29).

In 1956 two new supersonic fighters appeared: the MiG-21 (30), and the Su-7 and Su-9 (31) designed by P. Sukhoi, who had produced some fighters in the early 1940s and had not been heard from in the following decades. These two designs are geometrically similar (except that one has a delta wing and the other a swept wing), and both allow essentially the same maximum speed (Mach1.5) and ceiling (maximum altitude at which level flight is possible: 65,000 ft.). The Sukhoi fighters have a range of over 1,000 miles; the MiG-21 is smaller, and its range is 500 miles. Some 1,000 Su-7 aircraft were built, many for export. The MiG-21 was a particularly successful design: since 1958 more than 11,000 have been built, and many are still in service in twenty-three Soviet satellite, Middle Eastern, and Asian air forces. The MiG-21 is produced under license in India; it is the only Soviet military aircraft assembled in plants outside "the Soviet bloc" (32).

The integration of advanced electronics into Soviet air weapons was difficult at first (33). In the 1940s and early 1950s the Soviets had not developed their skills in airborne electronics and computer systems. The necessary precision production facilities were barely available on a small scale under the jurisdiction of the Communications Ministry and the Communications Command of the Armed Forces, and consequently they could not readily satisfy the needs of the Air Force or of the Ministry of Aircraft Production. The conservative tendencies of Soviet electronic designers also made them reluctant to shift from vacuum to transistor technology until the late 1950s. The first Soviet radar-equipped all-weather fighter, the Yak-25, was designed in 1951 and tested in 1952; it seems to have experienced serious difficulties during its flight-test period and did not go into service until 1956, at which time its performance was inadequate by comparison, for example, with that of the MiG-21. Yet it went into large-scale production (over 10,000 built) and remained in service up to the mid-sisties (34).

However, in the course of a visit to the Moscow Aviation Institute in November 1958, the Exchange Mission on Engineering Education of the American Society for Engineering Education found a large, well-equipped Aircraft Radio and Electronics Department there, and the NASA Apollo-Soyuz team (1972) found their Soviet colleagues quite sophisticated in space communications (35). Clearly, after 1955 Soviet aerospace engineers had learnt the importance of integrating electronics into their designs, and the institutional changes needed to assemble the new systems were well under way.

The U.S. Context and Fighter Development Program

The pattern of American fighter development differed dramatically from the Soviet pattern. While the 1929–40 depression severely disrupted the U.S. economy, it caused no essential damage to the infrastructure or to production facilities. In fact, the Roosevelt administration tried to stimulate the economy, in part, by undertaking large construction projects such as dams, electric power systems, highways, airports, and the like (36). World War II was fought on foreign soil and the United States became "the Arsenal of Democracy"; the resulting federal investment in new plants, particularly for the large-scale production of aircraft and engines, made the United States the world's leader in aeronautics. Between 1940 and 1945, American plants built 243,234 airframes and 812,615 engines. During the war the Army Air Corps built a worldwide set of fixed bases, with concrete runways and sophisticated machine shops and communication facilities, and developed a strategy geared to long-range bombing of industrial and military targets that culminated in the fire-bombing of German and Japanese cities and the dropping of atomic bombs on Hiroshima and Nagasaki (37).

The U.S. aircraft industry in 1945 was supported by an uninterrupted, substantial program of applied research carried out at governmental facilities (e.g., NACA, laboratories at Langley Field, Virginia, and Ames Field near San Jose, California, Wright Patterson AFB, etc.), at universities with government funds (e.g., Cornell, MIT, Caltech, etc.), and in industry, also largely with federal support (38). Industry, the military, and the academic community had developed a smooth working partnership that aimed to keep American airpower technically the world leader. This was the continuation of policies and procedures developed during the war by the Office of Scientific Research and Development (39), and it also reflected the views of Gen. H. H. Arnold, chief of staff of the Army Air Corps, and of his chief scientific advisor, Prof. Th. von Karman of Caltech (40).

One lesson learnt from the British combat experience with radar in

1940—44, and from the need to coordinate the night operations of large numbers of bombers over Germany in 1943—45, was the paramount importance of aircraft-ground communications and of aircraft tracking and identification. The development of electronic devices and systems to help carry out these tasks was a high-priority joint objective of the military services, the MIT Radiation Laboratory, the Bell Laboratories, and other facilities in 1941—45 (41), and was energetically continued after the war.

In particular, American planners and designers began to think of the aircraft as the key element of an air weapon system that included ground control elements, electronic sensors and comunications, and increasingly sophisticated airborne radar. As early as 1941, the Army Air Corps had negotiated a contract with John F. Northrop, who had left his position as a senior design engineer at the Douglas Aircraft Co. in 1938 to start his own company. The new Northrop Aircraft Company was to design and build a night fighter, the P-61 Black Widow, which was equipped with airborne radar and carried a crew of two (a pilot and a radar operator); the Black Widow went into service in late 1944 and saw limited action at the end of World War II. Some 920 P-61s were built between 1942 and 1948 (42). By contrast, the prototype of the first Soviet Yak-25 was first seen in 1952 and went into service in 1956 (43).

We now turn to the administration of American fighter development at the end of World War II and in the succeeding years. That process involved the interaction of four sets of players with autonomous sources of authority and distinct functions and priorities (44).

The designers were employed by competing private airframe and engine manufacturers interested in making a profit; this required a strategy that maximized the prospects of a long production run and often included substantial upgrading of a model during its service life.

The customer was a Command of the Air Force (or the Navy), which specified the desired performance of the new aircraft to fulfill overall defense needs and, if possible also to enhance the importance of the Command's own role. For example, the original specifications for the TFX fighter written at the direction of General F. F. Everest, commander of the Air Force Tactical Air Command (TAC), in 1958 required the prospective new fighter to have the speed and altitude performance needed to protect U.S. airspace from attacking long-range bombers — and also to have a range of over 1,000 miles with a payload sufficient to carry a nuclear weapon, so that TAC could have a share in the U.S. strategic mission along with the Strategic Air Command (45).

The third set of players was the Congress, which appropriated the funds needed both to develop a given aircraft program and to finance aircraft procurement services for the military; chairmen and senior members of the committees involved also kept a very critical eye out for the economic interests of their districts.

The fourth set of players — after the passage of the 1947 Armed Forces Unification Act — was the Office of the Secretary of Defense, who planned and coordinated overall defense strategy and recommended the allocation of funds for new weapons systems within bounded total resources. As the secretary's staff developed independent sources of data and the authority and the will to use them, their role became very important, especially during the long tenure of Robert MacNamara as secretary of Defense (1960—68) (46).

After the military had defined a requirement for a new weapon, several firms, or teams of firms (airframe, engine, electronics, etc.), would submit bid proposals and designs from which the military would select one — sometimes on the basis of flight evaluation of actual prototypes, but also especially after 1953, on the basis of paper studies, to speed up the process and lower the preliminary development costs, which were rapidly growing as the complexity of fighter aircraft increased.

The main factors considered in the selection of a new aircraft were performance and price. Performance includes speed, ceiling, and range as well as armament, safety features, and the electronic devices that make an aircraft part of an effective combat team. The airframes manufacturer was the lead contractor of the project; traditionally, the airframe was the centerpiece of the system to which the engine and other equipment were attached. Engines were built by specialized firms (since 1946, mainly General Electric Corporation and the Pratt & Whitney division of United Aircraft Corporation, and the airborne electronic components, as well as the entire electronic system of which the aircraft was a part, were designed and built by a variety of firms such as the Bell Laboratories, RCA, or TRW. The integration of the entire system was the joint responsibility of the lead contractor and the customer. It became an increasingly time-consuming task, especially when the new aircraft also included several major untested components.

Pricing represented an important problem. It was difficult to set the cost of a new aircraft before it was fully developed, and before the unforeseen difficulties (*unks*, for unknown factors) and the unpredictable ones (*unk-unks*, for unknown unknown factors) had been evaluated. Therefore, prices tended to increase as a new model was developed and tested (47). Many remedies for this situation were attempted (e.g., cost-plus-fee contracts, firm-price contracts, contracts with incentive and penalty clauses, etc.) with little success, because higher costs were often the consequences of modifications or design improvements introduced by mutual agreement of the manufacturer and the customer after the initial contract was signed.

Most aerospace companies depended on the military services for half or more of their income; they had a high stake in winning a substantial fraction of the competitions that they entered, and in gaining access to the eventual large sales available from a military follow-on production contract. This was particularly important since the transfer of skills from military to commercial operations and of know-how within a firm was difficult due to security restrictions and to differences between commercial and military standards. The cost of developing a fighter aircraft was reaching over half a billion dollars by 1956; entry into the military aircraft market, and survival there, was therefore both expensive and risky (48). Indeed, no new firm has entered the field of fighter design since 1940. Several firms have abandoned the field (e.g.: Ryan, Curtiss, Republic Corp. are no longer major military aircraft contractors; Lockheed has focused its efforts on other projects); others have merged with financially stronger competitors (e.g., Douglas Aircraft Co. and McDonnell Aircraft) or have become divisions of large conglomerates (Bell Aircraft is now part of Textron; Consolidated Vultee has been absorbed by General Dynamics; North American Aviation now belongs to Rockwell International). To keep their design teams intact, airframe manufacturers have increasingly resorted to the preparation of joint proposals with other airframe manufacturers (e.g., General Dynamics Grumman for the TFX), or with engine and electronic component makers. This procedure is also customary in Europe (e.g., English-French cooperation in the Concorde SST project; French-German-Italian-Spanish joint Airbus projects; the NATO common fighter project under English leadership).

The intensity of competition for air weapon system contracts was somewhat mitigated by an informal division of labor among aerospace companies. For example, the Boeing Company has held a dominant position in the design and production of long-range bombers, since B-17 Flying Fortress days (1934); on the other hand, Boeing has not received a contract to build a fighter since 1929, and its last attempt to compete for one (the TFX, 1958—62) was a well-publicized failure. North American Aviation Corporation, by contrast, has been a leading fighter designer since it built the highly successful P-51 during World War II; in particular, it was responsible for both the F-86 and the F-100, the leading subsonic and supersonic fighters of 1948—58. J. R. Kurth gives a general analysis of this pattern as an underlying feature of U.S. military air procurement (49).

The financial strain of competing for military contracts was less severe for the firms that supplied components for fighter aircraft. For example, the segment of the electronic communications and computer industry in the design of air weapon components consisted of a few very large, and many smaller, more-specialized firms driven by the commercial market imperative to put heavy stress on continual innovation. Their relation to the military market was complex. Some, such as AT&T Bell Laboratories, considered defense work a patriotic contribution in a time of need, but were reluctant to remain heavily involved in military projects over long time periods (50). Others started their existence as a creation of the military. TRW (originally Thompson-Ramo Woolridge) was set up by the Air Force to coordinate the development of ballistic missiles in the 1950s; having become large and successful during the sixties, TRW then diversified the applications of its technological knowledge to a variety of commercial fields. Electronic products and devices have a wider range of applications than fighter aircraft, and they often have low unit costs. The electronic firms therefore find operation in the commercial communications market easier and more natural than Boeing and Lockheed found entry into the field of commercial hydrofoil boats, or Boeing and Grumman into the manufacture of lightweight streetcars and buses.

Yet the fact remains that the system of infrequent competitions for the award of potentially very lucrative assignments compelled the airframe companies to live dangerously and to accept large technical and financial risks as a normal condition of their existence. In particular, the competitors were strongly motivated to produce a design that stretched the technical state-of-the-art. They occasionally took risks (or were very optimistic) about fulfilling promised performance and cost standards and schedules, especially when the competitions were paper competitions. This tendency was encouraged by the military services, whose officers felt that the second best in combat is not good enough. In fact, the contractors and the service negotiators had a joint interest in being optimistic about "buy in" prices for prototypes; if Congress financed the development and pilot phases of a particular weapon, they were likely to accept a higher price for the production run.

The case of the TFX fighter contract competition (1958–62) provides an interesting example of several aspects of American military aviation procurement policy (51). The TFX was at first to be a long-range high-performance aircraft designed to meet the Air Force Tactical Air Command requirements for a new fighter-bomber; but the secretary of defense suggested with increasing insistence after 1961 that its design be so arranged (e.g., by allowing the wings to "swing" and thus operate at several sweep angles) that it would also meet the Navy requirements for a new carrier-based patrol and fighter aircraft. The finalists in the competition for the contract to design and test the multipurpose TFX prototype (a paper competition) were the Fort Worth (Texas) division of the General Dynamics Corporation and Boeing Aircraft Corporation of Seattle, Washington.

A first argument arose, in the evaluation of the two final bids, over technical performance. Both proposals met the performance specifications set by the services. Boeing proposed to use a new engine that had not yet completed its certification tests and to make some major structural components out of titanium, which performed better than aluminum at high temperatures, but for which there was no fully established manufacturing technology. The flight performance promised for Boeing's design exceeded specifications; it was superior to the performance promised by General Dynamics, who were proposing to use an available tested engine and an aluminum structure. But the General Dynamics design would be cheaper and was likely to be available sooner and to operate more reliably in service. Both Air Force and Navy uniformed service evaluation boards voted for Boeing and very high performance; they were overruled by the civilian service secretaries and by the secretary of defense, who preferred acceptable performance with more predictable costs and early availability.

The second argument was over "commonality"; this was the term used by the Office of the Secretary of Defense to describe the requirement that the proposed aircraft actually meet the rather different technical requirements for both its Air Force long-range and its Navy missions. Both services were skeptical about commonality; in particular, the Navy requirement for carrier operations would force upon the aircraft a weight penalty due to the need for a strengthened landing train, which might jeopardize range and altitude ceiling performance; on the other hand, the size and the weight needed to achieve the Air Force performance requirements that the Navy did not find essential would make the aircraft unwieldy on all but the largest carriers. Boeing offered two versions of the TFX that shared most, but not all, features, while General Dynamics offered a single design. Again, the service evaluators, especially the Navy officers, voted for Boeing; the secretary of the Navy supported this point. The secretary of the Air Force and the secretary of defense chose the General Dynamics design, which promised lower initial price, lower maintenance costs, and greater flexibility in use — and was consistent with the principle of uniformity, an important criterion of American military procurement over the years.

Finally, because of the open disagreements between the professional officers of the services and the secretary of defense and his civilian staff, the award of the TFX contract was carefully reviewed by the Senate Permanent Investigating Subcommittee, chaired by Senator J.

McClellan (D. Ark.), at a series of public hearings (January—November 1963) that often pitted the Texas delegation and its allies from the South and the Southwest, including Chairman McClellan, against Senator Henry Jackson of Washington State and his supporters from the Northwest and the East.

In the event, the development contract was awarded to General Dynamics, who completed the TFX prototype on November 24, 1962; the resulting production aircraft, the F-111 fighter, entered service in the U.S. Air Force in 1967, after some technical difficulties with its "swing wing" had been adjusted. A total of 562 F-111s were built in 1963—85 (52). The F-111 has seen service in Vietnam and was used in the recent strike on Libya (1986); in these missions, it operated as a long-range unescorted fighter-bomber. The U.S. Navy bought seven TFX aircraft in the early 1960s and never put them into service.

In this institutional context so different from the Soviet context, we return to the history of the development of American jet fighters. The Bell P-59 jet fighter (1942), like some of the early Soviet models, was a traditional airframe carrying a jet engine (53). In 1943, the U.S. Air Corps asked Lockheed's exceptionally gifted designer, Kelly Johnson, to build a true jet fighter. The result was the F-80, an excellent aircraft designed and built in 143 days (54). The first version of the F-80 was built around the imported British de Haviland Goblin centrifugal-flow engine and flew in January 1944. Shortly thereafter, a more powerful, American centrifugal-flow engine, the J-33, became available; a prototype powered by that engine was tested in June 1946. In the F-80, American pilots and crews learned to fly and operate jets and served in combat at the beginning of the Korean War; a total of 1,713 F-80 fighters were built; the trainer version (T-33) of this aircraft, of which 5,700 were built, remained in service for twenty-five years.

Meanwhile, a competition to produce the next generation of jets led to the introduction of the Republic Aircraft F-84 (55), effective particularly in ground support missions, and the North American Aviation F-86 Sabre (56), which was the mainstay of the U.S. fighter command from 1951 to 1956 and remained in service well into the 1960s. Each of these aircraft was powered by the new axial-flow J-46 engine built by General Electric. The F-86 underwent a number of modifications (1951—56) that increased its weight and its versatility by the addition of new auxiliary equipment and the use of a more powerful engine — first an improved version of the J-46, and eventually a J-73 engine. Altogether, some 6,300 F-86s were built in several variants for the USAF, and over 2,000 more under license for NATO allies.

In the Korean War, the F-86 and the MiG-15 fought many air battles. The American pilots' reaction to the MiG-15 was somewhat comparable to the Navy pilots' reaction to the Japanese Zero fighter ten years earlier: a surprisingly good aircraft, lighter and more maneuverable than ours, but unable to withstand heavy American firepower. The Korean combat record indicates that the loss ratio heavily favored the F-86, because of either better pilot training or better equipment. Simulated combat between F-86s and captured MiG-15s flown by American pilots suggested that the F-86, with better gunsights, heavier armament, and a more rugged structure, would shoot down the MiG-15 when it was allowed to get into firing position — but when a well-trained test pilot flew the MiG-15, he was often able to outmaneuver and evade the F-86, particularly at high altitude (57).

In the early 1950s, the design of a supersonic fighter was considered urgent in the light of U.S. intelligence interpretation of Soviet technical achievements. The U.S. Air Force had sponsored research at NACA, MIT, and Douglas Aircraft Co. on applied transonic aerodynamics and on the control of maneuvering transonic aircraft. Several experimental aircraft had been built, including the X-1, in which Charles Yaeger exceeded the speed of sound in level flight in 1947. A major engineering task was to design operational supersonic fighters, and the competition for the "Century Series" called for the actual construction and extensive testing of several alternative models powered by the new J-57 and J-79 axial-flow engines, designed under Air Force contract by the Pratt & Whitney division of United Aircraft and by the General Electric Corporation, respectively. These new aircraft included the North American F-100 Super Sabre (58), a slimmed-down swept-wing fighter based on the experience of the F-86 Sabre; the Convair (later General Dynamics) F-102 (59), the first American delta-wing design; the Lockheed F-104 Starfighter (60), a light, stripped-down, very high-performance, straight-wing supersonic aircraft, somewhat hard to handle on take-off and landing; and the larger, twin-engined McDonell F-101 Voodoo (61).

The F-102 proved to be underpowered with the J-57 engine and did not meet all of its specifications; when a larger engine (J-75) became available, a modified version of the F-102 became the all-weather radar F-106 interceptor, of which some 300 were built (62). The F-104 was deemed marginal because of its tricky handling characteristics; only 180 were delivered to the U.S. Air Force. It was eventually purchased and used extensively by the West German Air Force and other NATO allies (2,700 were produced under license). The F-104's accident record over the years in Europe confirmed the judgment that it was tricky to handle. The F-100 was chosen to be produced in series for the U.S. Air Force (over 2,300 were manufactured), and it became the standard U.S. fighter in the period 1954—72.

Concluding Comments

The discussion of U.S. and Soviet fighter development leads to several conclusions. A national "style" — based on history, geographic conditions, and institutional and political constraints — affects both the final product and, more importantly, the process by which that product is achieved. Soviet designers, working in a political system that extracted high prices for mistakes (prison, etc.), took a conservative, step-by-step approach with a bias toward simple solutions that economize skilled manpower and avoid elaborate tooling and complex instrumentation. The separation between design and production, and a strategic doctrine emphasizing the need for a large defensive force in being, encouraged the early freezing of designs and the fairly frequent introduction of new models on a large scale. The close-knit community of Soviet airframe and engine designers tended to tailor their aircraft to overcome weaknesses in production technique by increased engine power. For technical as well as institutional reasons, sophisticated electronic equipment became a major component of the aircraft somewhat late in the period under study.

The general attitude of Soviet aeronautical engineers is well characterized by the following guidelines suggested by A. S. Yakovlev, the last survivor of the generation whose work is examined in this study:

1. Maximum simplicity for both reliability and production.

2. Evolutionary development to minimize both risk and impact on production.

3. Minimum need for field maintenance to insure the ability to operate independently of climate and conditions and the availability of field support (63).

The American contracting system, on the other hand, led to a bias toward sophistication in the design and manufacture of all components – often to the benefit of effectiveness, but sometimes leading to delay, excessive growth in cost, and difficult maintenance in the field. It also encouraged a marketing process that was potentially destabilizing to those airframe manufacturers who were primarily engaged in military work.

While the Soviets produced a total of 25,000—30,000 MiG-15s, MiG-17s, and MiG-19s in 1948—56, the United States and its allies produced some 10,000 comparable fighters, mostly several versions of F-86s in 1951—56; similarly, some 11,000 MiG-21s were built between 1956 and the 1970s, as compared to some 6,000 F-100s and F-104s and comparable American supersonic fighters. These numbers underscore the great importance of a large force-in-being to the Soviets. They also indicate that by continually upgrading an aircraft such as the F-86, the Americans obtained performance improvements comparable to those the Soviets obtained by introducing several new models; in fact, the F-86F differed as much from the F-86A or the F-86C as did the MiG-19 from the MiG-15.

A final word needs to be said about national style. As aerospace technology evolved, and as the Soviet Union repaired its war losses and developed a modern industrial base, did the characteristics observed in 1945—60 remain valid? A new generation of designers has replaced Tupolev, Yakovlev, Sukhoi, Mikoyan, Gurevich, and their contemporaries (64), but the distinction between the design function and production has remained in force. The personal risks of failure have in most cases been greatly softened, but the opportunities for major technological risk-taking still often depend on direct access to the leader and on his direct support.

N. S. Khrushchev personally involved himself in the planning of space operations in 1956—60. He provided considerable resources for designer S. P. Korolev's laboratories, design bureaus, and plants; but he also took a hand in selecting missions and landing sites and in scheduling launches. Both he and other senior officials, such as the commander of Strategic Missile Forces, Marshal M. I. Nedelin, kept close contact with the actual tests of the Vostok vehicle. In fact, the marshal died on October 23, 1960, along with several other senior officials, in the course of a launching accident. Korolev suffered a heart attack in December 1960; over medical objections and under party pressure, he continued his work at an intense pace and died in January 1966 (65).

The Soviet policy of "technical satisficing" still appears to guide the design of their sophisticated space vehicles, as it did their fighter aircraft. The Soviet space vehicles show considerable skill in design and sophistication in manufacture, and were highly praised by NASA staff in the close interaction that occurred during the Apollo-Soyuz project (1972). But some of the differences in emphasis between Soviet and American designs remain. For example, the Soviet manned cabins are filled with air at atmospheric pressure. Early U.S. cabins carried pure oxygen at 0.25 atmosphere; this allowed a more elegant, lighter structural design but accepted the risk of accidental explosion, which did occur in 1966 and cost the lives of three astronauts. Soviet boosters have on the whole been slightly larger than American ones, as their jet engines were in 1950—60.

The same emphasis on "satisficing" appears in the history of Soviet development of transport aircraft in 1955—80. At the end of World War II, both Western and Soviet civil airline systems underwent a very rapid expansion, which required both new ground facilities and new flying materiel (66). The traditional propeller-driven commercial transport aircraft (DC-6 and DC-7, Constellations, etc.) were becoming obsolete as a result of progress in engine technology. With few exceptions, airlines in the United States and in Western countries, led by Pan American Airways, replaced their propeller-driven equipment with pure-jet aircraft (Boeing 707 and Douglas DC-8 introduced in 1958) (67), when the use of Turboprop engines would have been initially safer and cheaper; only a few local flights and, for a short time, the British carrier (BOAC) attempted to rely on turboprop aircraft. Public taste for high performance (speed, and cabin noise reduction) and the policy of politically set fares, which shifted the locus of airline competition from price to performance and comfort, eliminated any possibility of wide use for turboprop-driven aircraft (68).

At the same time, the Soviet airline system (Aeroflot) introduced a family of rugged, simple turboprop aircraft of various sizes and ranges (An-12, An-24, Il-18, etc.) on its domestic flights and introduced the pure-jet Tu-104 on international lines, where Aeroflot competed with Western carriers (1957). The policy of playing safe and introducing innovations in small steps was followed here also; in the mid-1960s Aeroflot began to introduce a range of jet aircraft (Tu-124, Tu-134, Tu-154, Il-62,etc.) on its domestic lines, and now the task is nearly complete (69).

Both in the United States and in the USSR, the demand for military aircraft and for space vehicles is driven by performance and by political imperatives rather than by the market. Public managers in both countries try to use the incentives of competition among a few selected producers to stimulate needed technical improvements at critical stages of the supply process; but that competition occurs under very special controlled conditions that bear little resemblance to market competitions.

Yet the examples described in this study suggest that American and Soviet designers follow different styles. The gap between what American and Soviet technology can achieve at their best has substantially narrowed since 1960; but Soviet designers still operate on a narrower resource base than their American colleagues, and they have grown up in a context of different incentives and constraints.

In spite of some similarity in the military procurement process, American and Soviet designers approach their tasks in a spirit that reflects differences in the outlook of their societies. In the United States, the market mechanism encourages individual initiatives; a generally open abundant society is fascinated with technical cleverness, and it rewards success more notably than it penalizes failure. Even in military procurement, the prescriptive style in which specifications are written for new devices is modified into a continuous dialogue between the military customer and the engineer and his employer who propose to improve their product over its entire useful life.

In the USSR, technical objectives are set to meet specified social or military needs. Designers have a limited part in setting those objectives. They then work with bounded resources to fulfill their assignments within that set framework, and if the risk and the cost of failure have decreased since the 1930s, the rewards of success remain limited.

Notes

1. *Jane's All the World's Aircraft*, 1941 ed. (London: Jane's Publishing Co., 1941), p. 99c.
2. W. A. McDougall, *The Heavens and the Earth* (New York: Basic Books, 1985), p. 42; A. S. Yakovlev, *Fifty Years of Aircraft Construction* (Moscow: Nauka, 1968; NASA translation, 1970), p. 101.
3. *Jane's All the World's Aircraft* (1), 1945—46 ed., p. 39c. The Gloster Meteor first flew in 1943 and was used extensibly to protect London against German V-1 buzz bombs in 1944.
4. *Ibid.*, p. 141c. The Me-262 was designed in 1938 and a prototype was flight-tested in 1940, but large-scale production was delayed until 1944.
5. M. Sherwin, *A World Destroyed* (New York: Random House, 1973), esp. chaps. 6—7.
6. J. B. Rae, *Climb to Greatness: The American Aircraft Industry 1920—1960* (Cambridge, Mass.: MIT Press, 1968), chap. 1. See also G. R. Simonson (ed.), *The History of the American Aircraft Industry* (Cambridge, Mass.: MIT Press, 1968), esp. 45.
7. R. E. Stockwell, *Soviet Airpower* (New York: Pageant Press, 1956), pp. 80—82. See also A. Perry, *Russia's Rockets and Missiles* (New York: Doubleday, 1960), pp. 146—148.
8. N. Duke and E. Lanchberry, *The Sound Barrier — The Story of High Speed Flight*, 2nd ed. (London: Cassell, 1953).
9. Data derived from appropriate yearly editions of *Jane's All the World's Aircraft*; from E. Angelucci and P. Matricordi, *World Aircraft — Military 1945—1960* (New York: Rand McNally, 1978); and from Ph. de St. Croix (ed.), *Fighters* (London: Salamander Books, 1981).
10. A. G. Korol, *Soviet Education for Science and Technology* (New York: Wiley; Cambridge, Mass.: MIT Press, 1957), chap. 4. See also N. de Witt, *Education and Professional Employment in the USSR*, NSF Report nos. 61—40 (Washington, D.C., 1961), pp. 504—509.
11. For a description of the operation of Soviet aircraft plants in 1939—46 by Soviet workers and observers, see J. A. Parker (ed.), *The Soviet Aircraft Industry*, Soviet Planning Study Series, no. 4 (Chapel Hill, N.C.: University of North Carolina Press, 1955).

12. "ASEE Engineering Education Mission to the Soviet Union, Final Report, November 1958," *Journal of Engineering Education* **49** (1959), 839—911. See also Engineers Joint Council Report on "The Training, Placement, and Utilization of Engineers and Technicians in the Soviet Union" (New York, 1961).
13. V. I. Tikhomirov, *Organizatsyia Planirovanyie Samolotostroyitelnogo Predpryiatyia* [Organization and planning of aircraft enterprises] (Moscow: Oborngiz, 1957), p. 610; see also chaps. 3—5, 10, 15.
14. J. Berliner, *The Innovation Decision in Soviet Industry* (Cambridge, Mass.: MIT Press, 1976), esp. chap. 1.
15. W. Gunston, *Aircraft of the Soviet Union* (London: Osprey, 1983), p. 89.
16. McDougall, *op. cit.*, 1985 (2), pp. 38—39.
17. J. E. Oberg, *Red Star in Orbit* (New York: Random House, 1981), pp. 20—25. See also McDougall, *op. cit.*, 1985 (2), pp. 37—38.
18. A. Solzhenitsyn, *The First Circle*, trans. P. Whitney (New York: Harper & Row, 1968).
19. A. S. Yalovlev, *Rasskazy Aviakonstruktora* [Tales of an aircraft designer] (Moscow: Dietizdat, 1961; E. A. Tokayev, *Betrayal of an Ideal* (Bloomington: University of Indiana Press, 1956).
20. L. Trilling, *Soviet Education in Aeronautics: A Case Study* (Cambridge, Mass.: MIT Center for International Studies, 1955).
21. L. Trilling, "Soviet Technology," in *Encyclopedia on Russia and the Soviet Union*, ed. M. Florinsky and H. Schwartz (New York: Donat, 1961), pp. 557—560. See also L. Trilling, "Soviet Aeronautical Scientists: How They Work and Where They Publish," *Aerospace Engineering* **20** no. 7, (July 1961), 12.
22. Yakovlev, *op. cit.*, 1968 (2), pp. 100—106, esp. p. 101, presents the reminiscences of a participant. See also W. Green and G. Swanborough, *The Observer's Soviet Aircraft Directory* (London: F. Warner, 1975), pp. 55—60; this reference confirms Yakovlev's story but places the date of Stalin's decision to order a competition in March 1946. In view of the fact that the MiG-15 was flight-tested in late 1947, the earlier date appears more plausible.
23. Gunston, *op. cit.*, 1983 (15), pp. 174—175.
24. Sir Arnold Hall, "Some Comments on Current Aviation Topics," *Journal of Aeronautical Sciences* **24**, no. 3 (1957), 161—187.
25. Gunston, *op. cit.*, 1983 (15), pp. 176—178.
26. This expression, often encountered in Russian lore, refers to the episode in which Prince Potemkin, the governor of the Ukraine at the time of Empress Catherine I, built movable villages during an inspection trip of the empress; these would be assembled in the morning at the site of her Majesty's visit and disassembled in the evening, to be reassembled the next morning at another inspection site.
27. Tikhomirov, *op. cit.*, 1957 (13), Part III, Ch. 10.
28. W. Sweetman, *Soviet Military Aircraft* (London: Hamlyn Aerospace, 1981), p. 81.
29. *Ibid.*, pp. 142—143.
30. *Ibid.*, pp. 87—99.
31. *Ibid.*, pp. 142—150.
32. Gunston, *op. cit.*, 1983 (15), pp. 180—184.
33. See P. J. Murphy (ed.), *The Soviet Air Forces* (Jefferson, N.C., and London: McFairland, 1984), esp. pp. 78—79.

34. Sweetman *op. cit.*, 1981 (28), p. 192; Green and Swanborough, *op. cit.*, 1975 (22), pp. 71—75.
35. E. C. Ezell and L. N. Ezell, *The Partnership — A History of the Apollo Soyuz Project*, NASA History Series (Washington, D.C.: NASA, 1978).
36. E. E. Freudenthal, "The Aviation Business in the 1930's," in Simonson, *op. cit.*, 1968 (6), pp. 73—118.
37. I. B. Holley, Jr., *Buying Aircraft: Material Procurement for the Army Air Forces* (Washington, D.C.: U.S. Army, Office of the Chief of Military History, 1964), pp. 576—582.
38. A. Roland, *Model Research* (Washington, D.C.: NASA, 1985), vol. 2., esp. chaps. 1—6.
39. G. Pursell, "Science Agencies in World War II: The OSRD and its Challenges," in N. Reingold (ed.), *The Sciences in the American Context: New Perspectives* (Washington, D.C.: Smithsonian, 1979), pp. 359—378. See also Sherwin, *op. cit.*, 1973 (5), pp. 40—67.
40. P. A. Hanle, *Bringing Aerodynamics to America* (Cambridge, Mass.: MIT Press, 1982), pp. 135—138.
41. D. J. Kevles, *The Physicists* (New York: Random House, 1971), pp. 303—308, 341—348.
42. *Jane's All the World's Aircraft* (1), 1946 ed., pp. See also L. S. Jones, *U.S. Fighters* (Fallbrook, Calif.: Aero Publishers, 1975), p. 163.
43. Sweetman, *op. cit.*, 1981 (28), p. 192.
44. Holley, *op. cit.*, 1964 (37).
45. R. J. Art, *The TFX Decision* (Boston: Little, Brown, 1968), esp. pp. 15—27.
46. A. C. Enthoven and K. W. Smith, *How Much Is Enough?* (New York: Harper & Row, Colophon Books, 1971).
47. H. K. Hebeler, "Pricing Invention," MIT Seminar, October 16, 1969. Mr. Hebeler was then a senior official at Boeing Aircraft Corporation.
48. J. S. Gansler, *The Defense Industry* (Cambridge, Mass.: MIT Press, 1980), esp. pp. 42—48, 148—151, 170—184.
49. J. R. Kurth, "A Widening Gyre: The Logic of American Weapons Procurement," *Public Policy* (Winter 1971), 373—404.
50. See M. D. Fagen (ed.), *A History of Engineering in the Bell System*, vol. 2, *National Service in War and Peace (1925—1975)* esp. p. 11.
51. Art, *op. cit.*, 1968 (45); the account given here is largely based on Art's study, esp. chap. 5.
52. Jones, *op. cit.*, 1975 (42), pp. 297—300. Also *Jane's All the World's Aircraft* (1), 1982 ed.
53. Jones, *op. cit.*, 1975 (42), pp. 152—154.
54. *Ibid.*, pp. 202—204.
55. *Ibid.*, pp. 214—216.
56. *Ibid.*, pp. 235—241.
57. *Encyclopedia of U.S. Air Force Aircraft and Missile Systems* vol. 1, *Post World War II Fighters 1945—1973*, ed. M. S. Knaack (Washington, D.C.: USAF Historical Advisory Commission, 1978), pp. 54—55, 58—59.
58. Jones, *op. cit.*, 1975 (42), pp. 265—268.
59. *Ibid.*, pp. 272—274.

60. *Ibid.*, pp. 278—281. See also *Jane's All the World's Aircraft* (1), 1976 ed.
61. Jones, *op. cit.*, 1975 (42), pp. 269—271; a total of 760 F-101 aircraft were built.
62. *Ibid.*, pp. 285—287.
63. Murphy, *op. cit.*, 1984 (33), p. 78, quotes an essay by A. S. Yakovlev.
64. U. Albrecht, "Continuity and Change in Soviet Design Leadership, the Case of the Aircraft Industry," unpublished note, Free University of Berlin, 1986.
65. McDougall, *op. cit.*, 1985 (2), pp. 241—244, 295—297.
66. H. McDonald, *Aeroflot — Soviet Air Transport since 1923* (London: Putnam, 1975).
67. R. E. G. Davies, *Airlines of the United States since 1914* (Washington, D.C.: Smithsonian, 1982), p. 508.
68. D. C. Mowery and N. Rosenberg, "Government Policy and Innovation in the Commercial Aircraft Industry," unpublished paper, Stanford University, November 1980.

STELLAR-INERTIAL GUIDANCE: A STUDY IN THE SOCIOLOGY OF MILITARY TECHNOLOGY

DONALD MACKENZIE

University of Edinburgh

Introduction

To an outsider, the world of technology is opaque. It is so by design, not accident. Increasingly, successful technologies are metaphorically, and sometimes literally, black boxes: give them the necessary input, and the desired output appears, without the user's intervention or understanding.

Nowhere is this more true than in nuclear weaponry. Secrecy obviously contributes to opacity, but so too does the desire to make the procedures for the control and use of nuclear weapons as automatic as possible. Not only are systems designed to be proof against their unauthorized use, they are also typically designed so that those immediately responsible for using them are spared the knowledge of the actual targets against which they are directed — at least in the case of strategic missiles.

The opacity of the technical is to be found too in most social writing about nuclear weapons. There, an implicit distinction is normally found between *goals* (whether these be the goals of a state, a government bureaucracy, a branch of the armed services, or whatever), which are seen as appropriate subject-matter for the social scientists, and *technical means*, which are not. Thus the social scientist can debate whether more accurate missiles are desirable, or why country A or group X wants them while country B or group Y does not; but how this greater accuracy is to be achieved would normally be taken as falling outside the purview of social science.

This paper, by contrast, will examine how far one can get by following a different approach, one that does not adhere to a rigid distinction

E. Mendelsohn, M. R. Smith and P. Weingart (eds.), Science, Technology and the Military, Volume XII, 1988, 187—241.

between "means" and "ends." Central to the analysis will be a "technical means," one indeed designed to enhance missile accuracy: stellar-inertial guidance. Most strategic missiles are guided by a self-contained "pure inertial" system. In stellar-inertial guidance this is supplemented by a star sensor, which can take a "fix" (or fixes) on one or more stars while the missile is in flight.

Stellar-inertial guidance is used in both Soviet and American submarine-launched ballistic missiles. Although early missiles of this kind were guided by pure inertial systems, the United States began to deploy stellar-inertial guidance when the Trident I system entered service in 1979. Unusually, however, the Soviet Union was ahead: the SS-N-8 missile, first deployed in 1973, carries a stellar-inertial system, as does the SS-N-18 deployed in 1978 (1). (See table 1 for more details.)

The potential significance of stellar-inertial guidance is considerable. For close to three decades, strategic theorists have tended to see the nuclear-missile submarine as the optimum retaliatory weapon — invulnerable under the ocean, but lacking the accuracy needed to conduct a "first strike" against the other side's silos and command posts. If stellar-inertial guidance can help make submarine-launched missiles as accurate as their land-based counterparts, an important part of the strategic equation may be altered. Crucially, the "second generation" American stellar-inertial missile, Trident II, due for deployment in 1989, is intended to have unprecedented "hard target kill capability."

Despite the significance of stellar-inertial guidance, however, no history of it has ever been written, and few are the social scientists who have paid it any attention. For example, in a recent 500-page book on Trident, stellar-inertial guidance is taken as meriting only a few lines and not even an index entry (2). In the social sciences literature on nuclear weaponry I have been able to find only one serious, albeit brief, discussion of stellar-inertial guidance, a discussion that conforms largely to the above pattern. The analysis focuses on whether the goal of enhanced submarine-launched ballistic missile accuracy was desirable in the light of different strategic perspectives and bureaucratic interests; attitudes to stellar-inertial guidance are taken as more-or-less straightforwardly reflecting attitudes to this goal, with those who wanted greater accuracy wanting stellar-inertial guidance, and those opposing the one opposing the other (3).

TABLE 1
American and Soviet Submarine-Launched Ballistic-Missile Guidance History

American			Soviet		
Missile	First Deployed	Guidance	Missile	First Deployed	Guidance
Polaris A1	1960	Mk 1 inertial	SS-N-4 "Sark" [surface-launched]	1958 (?)	inertial
Polaris A2	1962	Mk 1 inertial			
Polaris A3	1964	Mk 2 inertial	SS-N-5 "Serb"	1963	inertial
Poseidon C3	1971	Mk 3 inertial [Mk 4 stellar-inertial option cancelled.]	SS-N-6	1968	inertial
Trident I C4	1979	Mk 5 stellar-inertial	SS-N-8	1973	stellar-inertial
Trident II D5	1989 [flight tests began Jan. 1987]	Mk 6 stellar-inertial	SS-N-17	not operationally deployed	inertial (?)
			SS-N-18	1978	stellar-inertial
			SS-N-20	1983	sources conflict
			SS-N-23	flight tests began 1984	?

Sources: Strategic Systems Project Office, *FBM Facts, FBM Chronology* (Washington, D.C.: Navy Department, 1978, 1982); *World Weapon Database. Vol. 1, Soviet Missiles* (Barton Wright) (Lexington, Mass.: Lexington Books, 1986); interviews.

To write a history centered on a technical means obviously requires both data and at least a rough framework within which to analyze them. Although I have made use of such documentary evidence as was available to me, much of what follows is based on interviews conducted in the United States between 1984 and 1987 as a part of wider research on the history of inertial guidance and navigation. Around twenty of those interviewed had important involvements with stellar-inertial guidance — as developers, skeptics, military or political decision-makers, or members of the intelligence community scrutinizing guidance technology in the Soviet Union.

Clearly the information available to me on developments in the United States is much richer than that on developments in the Soviet Union. What follows is thus essentially a history of stellar-inertial guidance in the United States. The Soviet case is, however, interesting as a comparative backdrop. The other countries of possible interest are France and Britain. It is certainly conceivable that the French submarine-launched ballistic missile program may be moving towards stellar-inertial guidance, but I have not definite information that this is so. Britain is due to purchase Trident II missiles (minus their warheads) from the United States; this is, however, a "black box" purchase, and there has been, as far as I am aware, no British input into guidance system design or the decision to equip Trident with stellar-inertial guidance.

The framework of analysis I shall use to make sense of these primarily American data is drawn from recent work on technology that has been inspired by the "new" sociology of scientific knowledge. One useful concept to be found there is "stabilization" (4). Stabilization of a scientific fact is the process through which statements move from being potentially contentious claims to being unproblematically accepted features of the world — and, indeed, eventually cease to be statements at all, and become incorporated into routine procedures and "black box" scientific instruments (5).

"Stabilization" of a technology, as I shall use the term here, also involves its becoming a black box, a necessary one with reliable characteristics. It involves the banishment, at least amongst those with the power to influence the technology and its fate, of two key potential ambiguities. The first is what the technology should consist in — what the contents of the future black box should be. In other words, a sufficient degree of consensus about "how to design" the technology should emerge. If it does not, the black box will never be wholly shut. People will constantly be seeking to reach inside and rearrange things. A graphic example of what this can mean is the history of MX, where chronic disagreement about the best basing mode played (and may continue to play) a major role in the troubles faced by the system.

The second potential ambiguity concerns the characteristics of the technology. Those whose opinions count must be convinced that the technology "works" (6) and "is needed." That is to say, they must be brought to believe that the technology reliably transforms its inputs into the desired outputs; that a means of doing this is beneficial, necessary, or at least unavoidable; and that no better way of doing it currently exists.

These two aspects of stabilization are inseparable. For example, if it is to be succesful, the process of design must be shaped by what it will mean for the technology that is its outcome to "work" and "be needed." This will involve taking into account not just the "physical" but also the "social" (economic, political, organizational, cultural) environment in which the technology will have to function. In relatively unusual but highly significant cases it may involve seeking to change that environment: new "needs" may be created, unfavorable economic circumstances transformed, laws altered, physical landscapes modified (7).

Reporting on social science research on nuclear weapons "from the technology outwards" necessarily means that this paper will involve "technical detail." While I will attempt to be as brief, simple, and clear as possible, discussion of detail is central to the enterprise — for detail is what the whole process of stabilization turns upon (8). The "social shaping of technology" (9) is not only to be found at the "macro level" of, say, societal preferences for nuclear or alternative energy sources; it is also to be found in the detailed steps and decisions of design and development. And the outcome of these can be wholly crucial to whether a technology will "work" and what its "social meaning" and "social effects" will be (10).

Failure to Stabilize: The Stellar-Inertial Navigation of Cruise Missiles and Aircraft

The use of star fixes is of course a navigational technique of great antiquity, and one that by 1939 had been adapted to the needs of long-range aviation (11). So it was scarcely surprising that when automatic navigation was being investigated in the post-1945 United States it seemed natural to employ star sightings there too. The technology of photoelectric detection had received a major boost during the Second World War (12), and a star tracker that could take fixes on a star automatically, even in daytime flight, seemed a feasible device to construct.

Star sighting could also correct for a major deficiency of a rival navigation system, inertial navigation. Inertial navigation is wholly self-contained, not relying on external inputs such as radio or radar signals, which wartime experience had shown to be potentially vulnerable to enemy jamming or deception. By 1945 no one had actually built a workable inertial navigator, but researchers in the Soviet Union, United States, and, especially, Germany and Austria had at least begun to investigate how such a self-contained system could be built. The German V-2 ballistic missile, though not inertially *navigated* in the sense that it was not required to "know where it was," was equipped with a limited form of inertial *guidance*: an "integrating accelerometer" sensed the missile's acceleration, and integrated this to form a measure of the missile's velocity, indicating when the missile had reached the required velocity and the rocket engine should be shut off (13).

As work progressed on inertial navigation after 1945, with the United States now playing a leading role, a "paradigm" design for an inertial navigation or guidance system emerged (14). In it, three accelerometers measured acceleration along each of three axes. These were in turn held in known orientation by being mounted on a "stable platform." The orientation of the platform was maintained by gyroscopes, also mounted on it, which detected any rotation of the platform and corrected it by feedback signal to servomotors controlling the platform. Figure 1 shows the outline of such an inertial navigation system. Although the proponents of inertial navigation faced skeptics who argued on a priori grounds that wholly self-contained navigation was impossible, they established early on that in principle such a system could work — provided, as we shall see, it were given certain initial information (15).

Principle and practice were, however, different. The dominant practical problem was making the basic sensors accurate enough for meaningful navigation. In the immediate postwar years, the focus of attention in the United States was not the ballistic missile — its

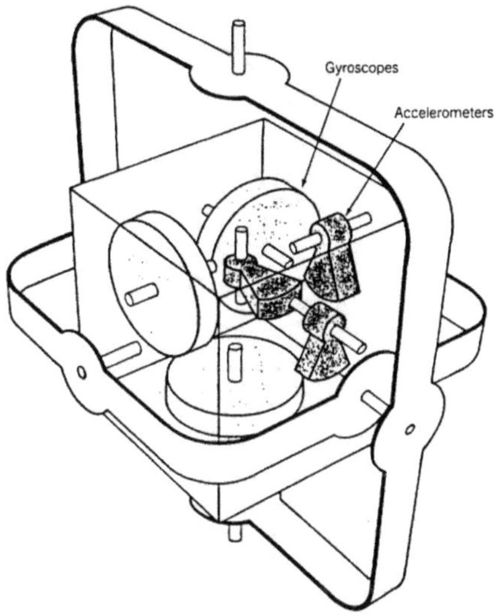

Fig. 1. 'Stable platform' configuration of inertial navigation system (highly schematic). (Source: Kosta Tsipis, "The Accuracy of Strategic Missles," *Scientific American 233* (1) (July 1975), 14—23, copyright © 1975 by Scientific American, Inc. All rights reserved.)

feasibility as an intercontinental weapon was influentially doubted (16) — but the bomber, or its automatic analogue, the long-range cruise missile. Flying probably at subsonic speeds, a bomber or cruise missile might take as long as ten hours to travel from the United States to its target in the Soviet Union. If the gyroscopes were to "drift" significantly — even by as little as hundredth of a degree per hour — a final error an order of magnitude bigger than the one-mile error desired by the Air Force could be produced. Even inertial navigation's most influential spokesperson, MIT professor Charles Stark Draper, had to admit that getting gyroscopes much better than a hundreth of a degree per hour was beyond the near-term state of the art, while other doubted that even a hundreth of a degree was feasible (17).

The use of a star sensor offered a way of circumventing this "reverse salient" (18). Gyroscope drift caused errors in orientation of the stable platform that tended to get worse as time went on. If, periodically, errors in orientation could be corrected, then system accuracy could be greatly improved. Taking "star fixes" offered precisely such a correction.

Figure 2 shows the kind of stellar-inertial system that was designed

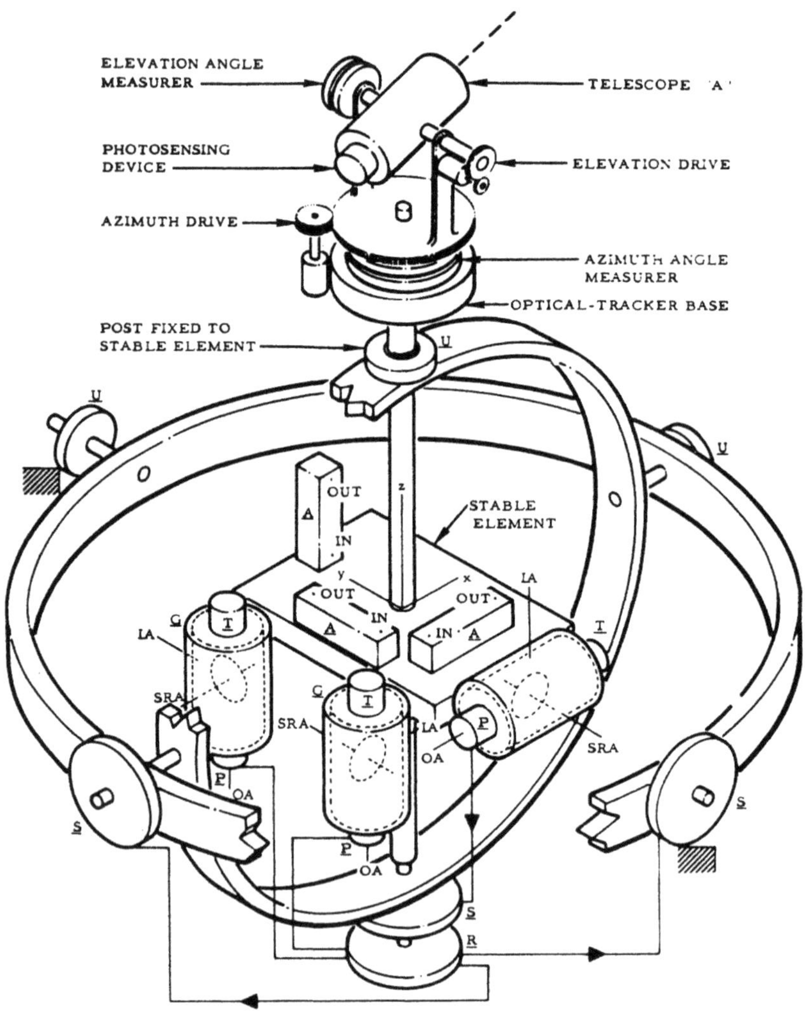

A ACCELEROMETERS SENSING ALONG x, y, z AXES

G GYROS STABILIZING ABOUT x, y, z AXES

P GYRO OUTPUT-AXIS PICKOFFS

T GYRO OUTPUT-AXIS TORQUERS

S GIMBAL SERVO MOTORS

R RESOLVER FOR GYRO PICKOFF

U GIMBAL ANGLE (ATTITUDE) PICKOFFS

Fig. 2. Stellar-inertial navigation system of the type designed in 1950s for long-range cruise missiles (Source: J. M. Slater, *Inertial Guidance Sensors*, New York: Reinhold, 1964, p. 204).

in the 1950s to do this. The bottom portion of the diagram, though more detailed than our very schematic figure 1, is a paradigm inertial navigator. Attached to this navigator is a star sensor, which has two degree of freedom with respect to the stable platform: it can turn relative to it in both azimuth (i.e., orientation in the horizontal plane) and elevation (orientation in a vertical plane). It can be pointed towards the predicted azimuth and elevation of a given star, and will either measure the difference between its predicted and actual azimuth and elevation, or "lock onto" the star and "track" it, continuously measuring its azimuth and elevation.

One assumption that was made about such a system is of considerable importance in our subsequent discussion – namely, that sightings on *two stars* were necessary, in order to correct fully for errors in platform orientation. A single star sighting provides only two pieces of information (see figure 3), while error in platform orientation has three components (i.e., rotations around three mutually perpendicular axes). As a 1962 textbook put it: "since a star tracker cannot detect angular errors about the line-of-sight, in order to correct all three axes of the platform, at least two stars must be tracked" (19).

Despite the advantages that stellar-inertial guidance was perceived as offering, stabilization of it as *the* means of long-range automatic navigation did not occur. Differences of opinion on design questions were still evident after a decade's work (20). In part, these reflected a basic

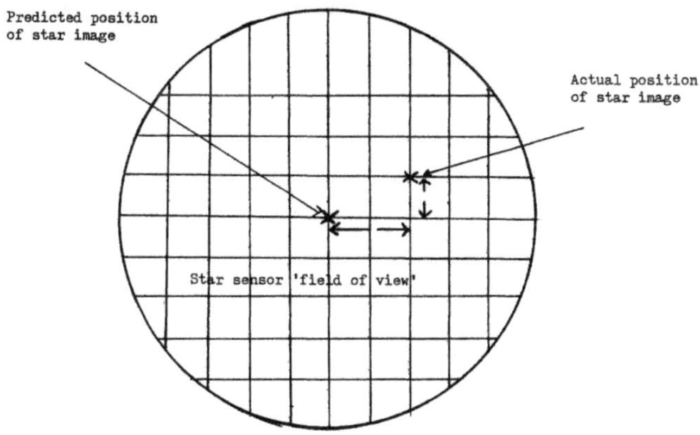

Fig. 3. The two pieces of information yielded by a single star sighting.

tension over the relative roles of the "stellar" and the "inertial" in stellar-inertial guidance. One design philosphy, most evident at the Northrop Aircraft Corporation, placed heavy reliance on the information from star tracking. A diametrically opposed philosophy, associated most strongly with Charles Stark Draper's MIT Instrumentation Laboratory, regarded the star tracker as a temporarily necessary, but messy and inelegant, "fix." Rather than circumvent the "reverse salient" of inadequate gyros, Draper's group preferred to attack it directly, refining basic inertial components through several generations until they no longer limited the performance of inertial navigators. A third position, found at the Autonetics Divisions of the North American Aviation Corporation, certainly saw supplementing inertial navigation with star tracking as attractive, but noted that features of the environment in which such a system would have to operate meant that the star tracker did not wholly remove the problem of gyro performance:

> In order to avoid operational restrictions, any system requiring stellar information must be able to start under overcast and continue to work during periods when star visibility may be interrupted by high altitude clouds during flight. It must also be able to work during daylight.
>
> The brightest astronomical objects (Sun, Moon, and planets) are generally undesirable for automatic systems because of their size and relatively complex apparent motion. Also, since they all lie within about 8 deg. of the plane of the ecliptic they may not be visible from polar regions, so operational problems in such regions would also rule them out.
>
> In order to keep oscillatory errors . . . from building up too much during periods of overcast and to carry along satisfactory information during such periods, *it is essential that any stellar-inertial system have high quality gyros*. This in turn lets the star trackers use a small angular field which, with present sensing techniques, helps provide daytime operation (21).

Central to the failure of stabilization of stellar-inertial guidance were the fluctuating fortunes of the weapon system for which it was primarily designed, the long-range nuclear-armed cruise missile. As noted above, this weapon initially seemed to many in the United States a much more feasible proposition than the intercontinental ballistic missile. A pilotless aircraft (which a cruise missile essentially was) seemed, especially to Air Force officers, a more familiar and credible prospect than a weapon that was more like a grotesquely scaled-up artillery round whose trajectory forced it to first leave and then reenter the earth's atmosphere. But by the mid-1950s the balance had shifted to the ballistic missile. The bureaucratic skill of the latter's proponents, intelligence assessments of Soviet successes with it, the increased destructive power (and proportionately smaller weight) of the new generation of hydrogen bombs, the fact that a warhead hurtling back from space seemed much harder to defend against than a pilotless plane flying for hours through the atmosphere — all these contributed to a massive growth in the ballistic-missile programs and a marginalization of the cruise missile (22).

For stellar-inertial guidance this was bad news. Taking star sights from an aircraft or its equivalent was one thing; taking them from a ballistic missile, quite another. The feasibility of doing so was easily dismissed out of hand, as in a 1958 article on guidance for the early Titan I intercontinental ballistic missile: "the inability to use stellar . . . references in a ballistic missile, as can be done in an aircraft or winged missile . . . greatly increases the accuracy requirements for ballistic missile acceleration sensors" (23). Even proponents of stellar-inertial guidance omitted the ballistic missile from the list of systems on which it might be used (24).

One intercontinental cruise missile program did go as far as deployment, but this was a mixed blessing for stellar-inertial guidance, as the missile was one that developed an unenviable reputation for unreliability and inaccuracy. Owing its name to Northrop's president's knowledge of the pages of Lewis Carrol, the Snark missile was deployed at Presque Isle, Maine, in May 1959. Although the program dated from 1946, even at the time of deployment the missile was dogged by persistent test failures. Of the last ten test flights, for example, only one was classed as successful:

> Two of the remaining nine broke up in flight, four were dumped after 2500 miles cruise, two had disabling electrical system defects, and one missed Ascension Island by some 420 miles after all the guidance system but the gyro controls had gone out of operation (25).

After a mere two years of deployment, Snark was decommissioned by the Kennedy administration in June 1961. It has, however, been outlived by a joke, repeated to this day by missile engineers, that because of the missile's many failures the waters off its Florida test base are "Snark infested."

Gradually, between the 1940s and 1960s, "common sense" about long-range guidance or navigation shifted. The beginning of the period was dominated by "the assumption that some guidance system that was an extension of autopilot and autonavigator experience would be 'easier' to develop than a closed-loop inertial trajectory system" (26). By the 1960s, however, the "pure inertial" solution had stabilized. The "in principle" arguments against it had been defeated; the many material and human problems in the ultrasensitive production processes it required had been at least partially overcome; the military had largely been convinced that its freedom from external input was a military advantage (even a military necessity). Draper's group had never needed convincing of the advantages of pure inertial navigation. North American Aviation, too, gradually abandoned its interest in star trackers. As a wider industry developed, and new markets opened up (first for tactical aircraft navigation, then for transoceanic civil navigation), there was little doubt that the pure inertial, rather than stellar-inertial, solution would be adopted. True, there were situations where economics prohibited the use of the highest-quality inertial sensors, or where other specific reasons led to the desire for additional sources of information to supplement inertial data. But neither in aircraft navigation, nor in the new generation of cruise missiles, did it become normal to seek that additional information from star sightings.

This situation is, however, the result of *decision*, not simply of "nature" or an impersonal "logic" of technology. One major organization still stands opposed to this consensus against cruise-missile and aircraft stellar-inertial guidance and navigation: the Northrop Corporation. For Northrop, the difficulties faced in the stellar-inertial guidance of Snark were a passing phase that has been overcome. The company has continued to develop, promote, and advertise stellar-inertial cruise and aircraft systems, arguing that no other technology combines high accuracy after long flight times with an absence of detectable emissions or dependence on jammable radio aids. Northrop's NAS-26 "precision astroinertial navigation system" is quoted as being capable of 1,000-

foot circular error probable after ten hours' flight — which would make it some five times more accurate than the most accurate operational pure inertial aircraft navigation system (27).

Although such systems are argued by Northrop "to set the standard for precision in cruise navigation systems" (28), they have not, to my knowledge, been adopted in deployed (rather than experimental) cruise missiles, nor in the major American strategic bomber programs. Such applications as they have found have been in "black" (classified) programs (29). One press report indicates the use of a Northrop stellar-inertial system in the SR-71 "Black Bird" reconnaissance aircraft program (30). The ultra-high security of black programs, by comparison with mainstream missile or military aircraft work, is, however, a major barrier to the adoption of technology from them in other areas. Even their existence as part of a black program cannot openly be disclosed, much less their success or performance levels in use. The flow of personnel, documents, and hardware is restricted even more stringently than in most military work, and work is often performed in facilities that are physically divided off from the rest of a company's plants. The SR-71 navigation work, for example, was reportedly performed in an unmarked Northrop facility in Hawthorne, California (31).

So if stellar-inertial guidance and navigation has stabilized in the aircraft and cruise-missile field, it would appear to have been under very particular conditions that kept this stabilization quite local — restricted to one corporation and to programs that cannot openly be discussed, even in military circles. Paradoxically, though, a much greater degree of stabilization of the technology was achieved in the one field for which it was initially dismissed as infeasible: ballistic-missile guidance. It is to this that we now turn.

The "Mobile Political Missile" and the Origins of Ballistic-Missile Stellar-Inertial Guidance

The Problem of Initial Conditions

By the early 1960s, then, pure inertial guidance in the United States had become a stabilized, "black box" technology. As with any black box, however, inertial guidance's appearance of wholly self-contained, autonomous operation was in part an illusion (32). To some extent this was because many early systems — especially aircraft navigators, but also some missile guidance systems, notably that of the American Minuteman II missile — were plagued with technical faults, and required a truly major maintenance and repair effort (33). But it was also because amongst the inputs needed by an inertial guidance system is information of a kind that is far from straightforward to supply. Obviously, it needs to "know" the location of its target, a matter not merely of identifying the latter (which became relatively easy with the advent in the early 1960s of photographic reconnaissance satellites) but also of pinpointing it in an accurate and consistent geodetic frame of reference. Less obviously, an inertial guidance system needs to be provided with "knowledge" of the gravitational field of the earth — a requirement that makes geophysics a subject of extreme military interest (34).

Most relevant to the history of stellar-inertial guidance, however, is a third requirement of an inertial system, which we can summarize as "initial conditions": before the launch of the missile carrying it, an inertial guidance system needs to know the *location* of its launch point, the *velocity* of its launch point, and its own *orientation*, all with respect to a known frame of refernce.

With missiles fired from fixed, land-based sites like silos, the first two can be determined well in advance of firing. Human error is known; a persistent bias in the results of the test-firing of American missiles was, for example, traced to a mistake in entering the survey figures locating the test silo. But launch position can in principle be determined with great accuracy, and from knowledge of that, together with knowledge of the rate of rotation of the earth, the velocity of the launch point in the chosen frame of reference can easily be determined — since a fixed silo is, obviously, at rest with respect to the earth.

An inertial guidance system can also be made to orient itself with respect to the local vertical — the process known as "erection." Orientation in the horizontal plane is more problematic, and early systems required skilled human intervention using an optical alignment system and external bench marks.

All in all, the problem of initial conditions in missiles fired from fixed, land-based sites has not seemed overwhelming, and even the delicate business of alignment has, at least in American missiles, gradually been automated (35). It therefore posed no threat to the black box nature of inertial guidance.

Skybolt and the MMRBM

What provided the opening for stellar-inertial guidance of ballistic missiles was an attempt to move away from fixed launch points — an attempt that immediately made the whole issue of initial conditions deeply problematic. Two possible missile systems were centrally involved (36). One was the Skybolt missile. Skybolt was an intensely ambitious scheme for a long-range ballistic missile that could be fired from an aircraft. The problem Skybolt was designed to solve was as much bureaucratic as strategic. Both the United States Air Force and the British Royal Air Force (which was also supposed to purchase Skybolt) remained deeply committed to the manned bomber as a weapon system. But with massive Soviet investment in air defenses, the capacity of such an aircraft to reach its target seemed increasingly in doubt; if a ballistic missile could be fired from a bomber, then this could be guaranteed. The bomber need not even enter Soviet air space, and so its physical survival (and therefore political survival) could be assured.

Stellar-inertial guidance was proposed for Skybolt, and a system was developed, again by Northrop. But Skybolt itself ran into deeper and deeper trouble, and was eventually canceled (37). Though it too was eventually unsuccessful, it was the other mobile missile system that is of greater relevance for our story: namely, the Mobile Medium-Range Ballistic Missile, or MMRBM.

Like Skybolt, the MMRBM was a technology designed to enter a difficult strategic and political environment. It was to be a European theater weapon, and reflected a growing sense in the late 1950s and early 1960s of the need to plan for a war that, though nuclear, might be in some sense limited. Unlike the existing fixed-base Thor and Jupiter missiles deployed in Britain, Italy, and Turkey, the MMRBM would be mobile, and thus protected against a Soviet preemptive attack. It was also, however, required to be highly accurate — because its "limited" use was anticipated — and ready to be fired at extremely short notice, perhaps as little as five minutes.

But the concept that a nuclear war could be limited was as far from universally accepted, and the territory on which the MMRBM would be deployed might not be welcoming. Saboteurs were one worry. More immediately, France was already on the path that was to take it to disengagement from the military structure of NATO, and the French government wished complete control over MMRBMs deployed on French soil, a proposal that would have been unacceptable in Washington (38).

Nor was Washington unified in its thinking on the MMRBM. It was to be an Air Force missile, but the Army was still smarting from the way the Air Force had defeated it in the battle for control over intermediate-range missiles, and could not be expected to welcome the MMRBM. The Department of Defense, particularly in the person of Assistant Secretary John H. Rubel, believed the Air Force to be too ambitious in the range and accuracy requirements being proposed for the MMRBM (39). There were accusations that the State Department was trying to block the MMRBM program by slowing negotiations with the European members of Nato for deployment rights, the suspicion being that the State Department saw the MMRBM as a rival to its pet project for a multilateral nuclear force (40).

This unstable political environment made the stabilization of the MMRBM as a weapon system extremely difficult:

> Since the decision was made last October [1961] to go ahead with the development of the missile, international and military politics have so affected attempts to complete a set of specifications on which to base requests for proposals to industry that some Pentagon officers have dubbed it the "Mobile Political Missile" (41).

Guiding the Mobile Political Missile

Designing the guidance system for the MMRBM was a process that could not be wholly insulated from these wider considerations. The mobility of the MMRBM immediately raised the issue of initial conditions, especially when combined with the demand of high accuracy. How was the system to "know" the exact location of its launch point, and how was it to be aligned?

One solution would have been to have a range of presurveyed launch locations, each equipped with a marker whose orientation was precisely known and which could be used for optical alignment of the MMRBM's guidance system. However,

> while this approach involves the least complexity for both missile and transporter, it runs the risk that the transporter might be blocked from reaching a launch site by bomb damage or sabotage. Additionally, it would require a permanent continuous guard at launch sites to assure their integrity and prevent sabotage of the azimuth reference benchmark (42).

Another would be to "black box" the problem of initial alignment within the transporter but not within the missile. For example, at the order to fire, the transporter could move to the nearest presurveyed launch point, but alignment could then be performed using a gyrocompass on the transporter. This would do away with vulnerable bench marks, but it was argued that it would compromise the requirement for rapid readiness to fire: "However, the north-seeking gyro would require 10—20 minutes to establish a reasonably accurate azimuth reference, and the transporter still must proceed to the nearest launch site to fire its missile" (43).

A further complication was that there were seen to be military and political advantages in altogether avoiding land-basing:

> Some Defense Department officials are inclined to believe that ship-based launch may ultimately prove the more feasible for a variety of reasons. One is the belief that water-based launching platforms would be less vulnerable to sabotage than transports operating over West European highways. Another is that water-based launchers would be less conspicuous and therefore less likely to provide a symbol which Communist or neutralist propagandists could use to create public opposition in some Western European countries (44).

Such a system could not rely on fixed external references for either initial position or azimuth. Similarly, an entirely free-ranging land-based system would also have to have its own means of ascertaining both of these. But was this possible without a radical deterioration of accuracy?

It was in the midst of this situation — one in which the "goals" to be met by an MMRBM guidance system were not stable and were strongly affected by its being "the mobile political missile" — that the proposal to guide the MMRBM by a stellar-inertial system emerged. Where the proposal first came from is not wholly clear. It may have been from a group at the AC Spark Plug Division of General Motors, which was one of the leading members of the nascent guidance industry. That group's leader, Hyman Shulman, and another prominent member, Martin Stevenson, had both worked at Northrop, and were well acquainted with the stellar-inertial work there.

The advantage that stellar-inertial guidance offered (if it worked – an issue to which we shall turn shortly) was that the problems of initial position and alignment could be black-boxed *within the missile*. This prospect was attractive enough to persuade the Air Force to issue three exploratory contracts under the Stellar Inertial Guidance System (STINGS) program, with a key role being played by a young technical officer, Robert Duffy.

With neither the technology nor the environment stable, a variety of different approaches were pursued under the STINGS program. The most radical approach was that of the United Aircraft Corporate Systems Center, which involved an effectively complete detachment of the missile guidance system from the politico-operational problems of initial conditions. It involved designing the missile guidance system so that it included *two* stellar sensors, which could take fixes on two stars simultaneously as the missile climbed into space:

> From measurements of azimuth and elevation angles of two stars relative to the horizon established by the stable platform, an on-board computer would calculate the missile's geographic position, compare this with the position it should have to hit its intended target and initiate steering signals to alter the missile's course as required (45).

With this system, it was argued, missiles could thus be fired from any point, and no azimuth reference, nor complex navigational equipment in the transporter, would be necessary.

Doubt about the feasibility of this approach to stellar-inertial guidance seems, however, to have been strong, and the two other STINGS contractors – AC Spark Plug, and the Kearfott Division of what was then General Precision Equipment Corporation (it is now part of the Singer group) – took more modest approaches. These are hard to define precisely, in part because neither stabilized, even locally, and in part too because the participants' memories are not wholly congruent with the available documentary evidence (chiefly contemporary reportage in *Aviation Week and Space Technology*). But some points seem reasonably clear.

First, unlike the radical version of stellar-inertial guidance, the point of these approaches was to use the stellar information to *correct accumulated errors* rather than directly to work out missile position. It was assumed that the system would have *some* information about both initial position and initial azimuth; any errors in the system's best estimate of these would translate into errors in angular orientation vis-à-vis the stars once the missile was in flight. (If one thinks of a frame of reference with its origin at the center of the earth, then latitude and longitude are angles with respect to this frame; that is why they too, as well as azimuth, translate into errors of angular orientation.) So differences between the actual and predicted positions of star images in the star sensor's field of view could be used to deduce errors in angular orientation, and thus to correct for errors in knowledge of initial position and azimuth.

Second, it was still assumed (for the reason outlined earlier) that for full correction it was necessary to have sightings on *two* stars. This could be done using a single star sensor, so long as it was mounted with freedom of movement with respect to the stable platform and a large enough "window" was provided — it could be pointed first at one star, then at the other.

Third, there was one scenario in which the use of a single star was envisaged. If the politico-military role of the MMRBM was such as to allow accurate knowledge of launch location through the use of presurveyed sites, then the error-correction problem was reduced to that of error in azimuth. One star was then adequate to correct for this, and a considerably simpler stellar-inertial system would therefore be possible (46).

The Beginnings of Stabilization

Testing Stellar-Inertial Guidance

Resolution of detailed design issues such as "one star or two" was difficult, because it appeared to require resolution of the more general political and military uncertainties surrounding the MMRBM. In actuality, no time was available for these resolutions. The overall failure of the MMRBM to stabilize meant that it was to remain merely a "political" missile, never a "real" one. After various fluctuations in its fortunes, the MMRBM program was finally canceled with effect from 31 August 1964 (47).

Interest in stellar-inertial guidance nevertheless remained. But if the technology was to get anywhere, doubts that it would work had to be stilled. Though, as we shall see, there were more sophisticated levels of doubt to be dealt with, a basic one concerned the very possibility of taking a star fix from a ballistic missile in flight, especially in daytime firings.

According to one interviewee, a whole series of experiments was attempted, using Polaris A2 and A3 missiles, Minuteman missiles, and high-altitude balloons. If his account is correct — no documentary evidence of these tests is available — it is of some interest because of the way it echoes the issues about scientific experiment and technological testing discussed in recent sociological literature (48).

What is of particular interest is how these experiments proved wholly ambiguous. There was a general problem of knowing whether or not a star sighting had successfully been made. An error in the star sighting could be a "real" error, or it could be that knowledge of the orientation of the stable platform on which the star sensor was mounted was incorrect. Even more seriously, if no star could be "seen" at all, what did this mean? That is what apparently happened in the early experiments, and it led those centrally responsible for the evaluation (it was reportedly contracted out to the Astronomy Department of MIT) to conclude that the experiments showed the infeasibility of stellar-inertial guidance. The brightness of the daytime sky was too great for the missile-mounted sensor to "acquire" a star.

An alternative interpretation of this "failure" was pushed by the proponents of stellar-inertial guidance. They argued that residue from the rocket engine might have become deposited on the "window," and sunlight reflecting on this layer was what had prevented the star's being seen. Initially, the Air Force could not be convinced that the problem was a remediable one of this sort. But eventually a further series of tests was agreed, in the Stellar Acquisition Flight Feasibility (STAFF) program, using Polaris A1 missiles being retired from active service and modified to carry a Kearfott-designed stellar-inertial system employing a Northrop star sensor.

The first such test was on 14 April 1965. Although range safety command aborted the flight when the missile veered off course just over a minute after takeoff, telemetry from the missile was accepted as showing that the star tracker had successfully locked onto the North Star, Polaris. This did not end doubts. The scientist responsible for the earlier evaluation noted that it was a night flight, and reportedly told a proponent of stellar-inertial guidance that he "would eat his hat" if the system succeeded in acquiring a star in daylight. However, the telemetry from the later daytime STAFF flights *was* taken as showing this. Ballistic-missile stellar-inertial guidance had finally passed its first feasibility test (49).

Unistar

But acceptance that a star fix could be made from a ballistic missile in flight in no way meant acceptance that stellar-inertial guidance was a useful technique, nor did it resolve the issue of how to design a stellar-inertial system. Interestingly, it was a theoretical argument — rather than an empirical test — that was to prove key in this respect.

By 1965 one particular organization had become paramount in pushing ballistic-missile stellar-inertial guidance: Kearfott. The United Aircraft and AC Spark Plug efforts had not found favor with either the Air Force or Navy. Kearfott had had to form a temporary alliance with Northrop to gain access to Northrop's expertise in star sensors. Its own attempts to develop a solid-state star sensor had run into difficulties, and worries about the brightness of the daytime sky had led to very demanding requirements for sensitivity in the star sensor. The Northrop star sensor used in the STAFF program employed a device rather like a television camera, known as a Vidicon tube (see figure 4). This technology Kearfott gradually came to master. But Kearfott's role in the stabilization of stellar-inertial guidance in the United States was, indeed, primarily conceptual.

The roots of this lie around the autumn of 1965, and central to it were two Kearfott engineers with strong statistical backgrounds, Irwin

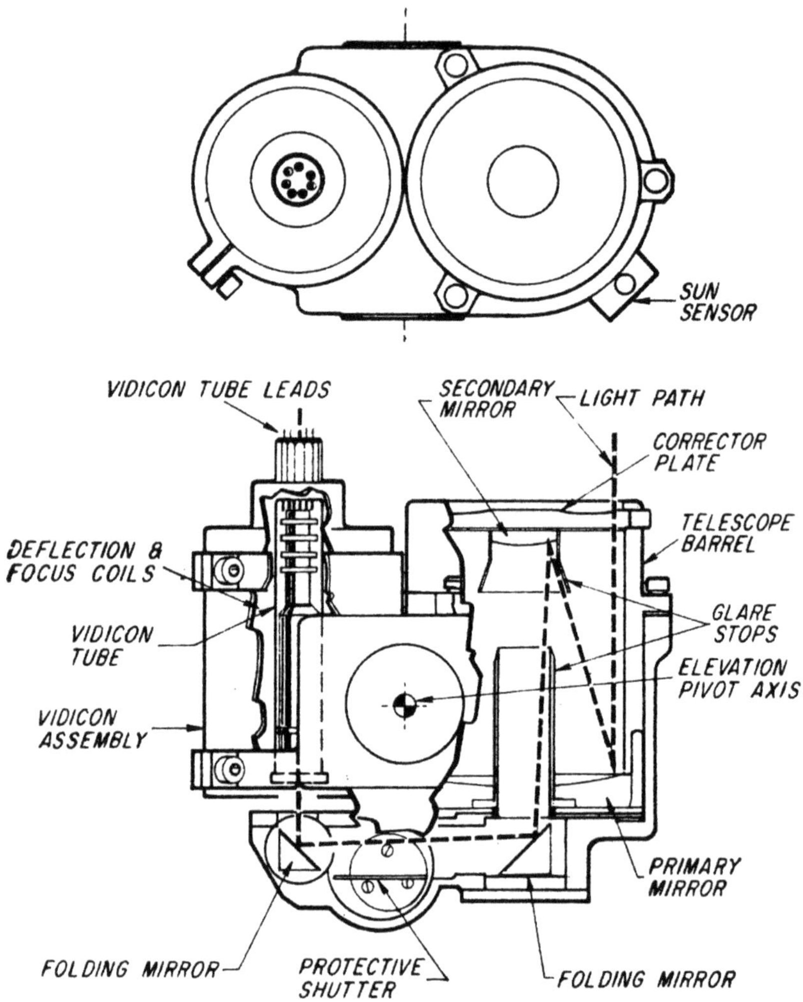

Fig. 4. Northrop (Nortronics) star sensor (Source: Barry Miller, 'Star tracker boosts MMRBM accuracy', *Aviation Week and Space Technology*, 27 April 1964, pp. 99–103).

Citron and Ted Mehlig. They were doing mathematical analyses of error behavior in stellar-inertial systems, and came to see a relationship in their models between the angular position of the stars on which fixes were taken and the final miss-distance of the missile. Gradually, they came to formulate a principle that undermined the previous assumption that two star sightings were necessary: if the first star were properly chosen, they suggested, then the second sighting would be redundant. As a later statement was to put it, the "Unistar principle" implied that "a single fix from an optimally located star will yield accuracy equivalent to a two-star fix" (50).

This was, patently, a counterintuitive argument, so it is worth following how it is justified in this later paper, one of the few publicly available detailed discussions of stellar-inertial guidance: Assume that the three errors in initial conditions for which the star fix has to correct are in latitude, longitude, and azimuth. On the face of it, it seems hard to imagine how the two pieces of information from a single star fix could correct three errors. Surely a fix on a second would help? But consider the situation of firing from the equator to a target located at the North Pole, with the star sensor prealigned on the star Polaris, located vertically over the Pole. Any error in initial latitude can be corected for, because it will lead to a determinate deviation in elevation between the predicted and actual images of Polaris. An error in azimuth will likewise be correctable from the deviation in azimuth of the position of Polaris. An error in longitude cannot be corrected for, however, because it has no effect on the location of the image of Polaris in the star sensor's field of view, as all lines of longitude converge at the Pole. But precisely because all lines of longitude converge there, an error in initial longitude does not cause a target miss. So a second star fix used to determine error in initial longitude would not enhance final accuracy.

This can be generalized to any target, argue the paper's authors, Stephen Rounds and George Marmar of Kearfott:

> Generally it is true that there is a direction in space about which attitude errors are uncorrelated with target miss errors. Therefore, if the single star to be sighted is the optimum star, the only axis about which the stellar sensor cannot measure attitude errors is exactly that axis for which it cannot make any correction. It follows that a second sighting to measure this third component of attitude error can provide no useful information. This is the Unistar Principle: multiple sightings of non-colinear stars cannot provide better accuracy than a single sighting of an optimum star (51).

Unistar and the Stabilization of Stellar-Inertial Guidance in the United States

Three kinds of doubts had to be overcome before the Unistar argument was accepted. The first was the basic matter of intuitions. Even though the proponents of Unistar could argue that the two cases were not analogous, the widespread understanding that in ordinary manual stellar navigation a single star fix was not adequate to yield position was clearly a barrier to be surmounted. In "insider" circles — guidance engineers, analysts, and the like — Unistar in fact seems to have won complete acceptance at the abstract level at which I have just presented it. But not all decision-makers were in this sense "insiders," and their contrary intuitions had to be reckoned with.

The second issue was the more practical one that there would not in general be a star of sufficient brightness in the optimum position. This was accepted by Kearfott, who went on to argue — and to expend considerable analytical energy seeking to prove — that the degradation in accuracy caused by the use of a not-quite-optimally-placed star would not be too serious. This seems to have been widely agreed, though it is worth noting that this kind of consideration led what is, to my knowledge, the first published statement of the Unistar principle to be significantly qualified:

> A single star-direction sighting will give only two components of data. Sighting on a second star at a substantial angle from the first star can give added data, but in a practical situation only one star properly chosen should provide nearly the full value that star sightings can give (52).

A third issue was that what was envisaged by Kearfott was not merely the use of a single star, but a single sighting of that star. The reason why this was of significance is that errors in platform orientation would not *simply* be caused by errors in knowledge of initial conditions. Drift in the gyros that were used to maintain the stable platform in known orientation would obviously also generate errors in platform orientation. Since errors with different causes necessitated different corrections, it was important to be able to separate them out.

Prior to Unistar, the systems envisaged at both Kearfott and Northrop envisaged taking repeated fixes on the same star (53). As Northrop MMRBM program manager Leonard Baker argued, "this would make it possible to determine drift of the platform gyros in dynamic conditions of acceleration" (54). Thus gyro drift and initial conditions errors could be separated out, and the proper correction applied.

With a single star fix, however, this could only be done with an a priori model. As Rounds and Marmar put it, "The attitude error observed by the stellar sensor results from a number of error sources and the flight computer software uses a priori knowledge of system statistics in order to make an optimum correction" (55). Confidence in this had to rest on a general confidence in this kind of statistical algorithm. While this was high amongst insiders, others sometimes took the attitude — expressed to me by one highly placed naval officer — that it was "Mexican arithmetic." More specifically, there had to be confidence in the a priori knowledge of system statistics — a point to which we shall return.

In the United States, none of these three forms of doubt was sufficient to prevent widespread acceptance of the Unistar argument. What this meant, and what it contributed to the stabilization of stellar-inertial guidance, was a considerable simplification of the mechanical design of the system. With only *one* star to point at *once*, there was no need to give the star sensor freedom to move with respect to the stable platform. It could be mounted integrally to it, and prealigned before launch so that it would point to the predicted position of the chosen star at the appropriate point in time. Not only was this an overall reduction in complexity, it made it easier to design the system so that its "line of sight" would not be distorted by the acceleration and vibrations of missile flight. The stable, paradigm American design of a ballistic-missile stellar-inertial system thus emerged; it is shown in figure 5.

Also important in this aspect of the stabilization of stellar-inertial guidance was a change in the circumstances under which the star fix had to be taken. Intuitively, rocket boost had never seemed the optimum time for it, but taking the fix while on the ground was ruled out: "Shooting the stars before launch is risky because of the chances of foul weather obscuring the star, especially in certain European seasons

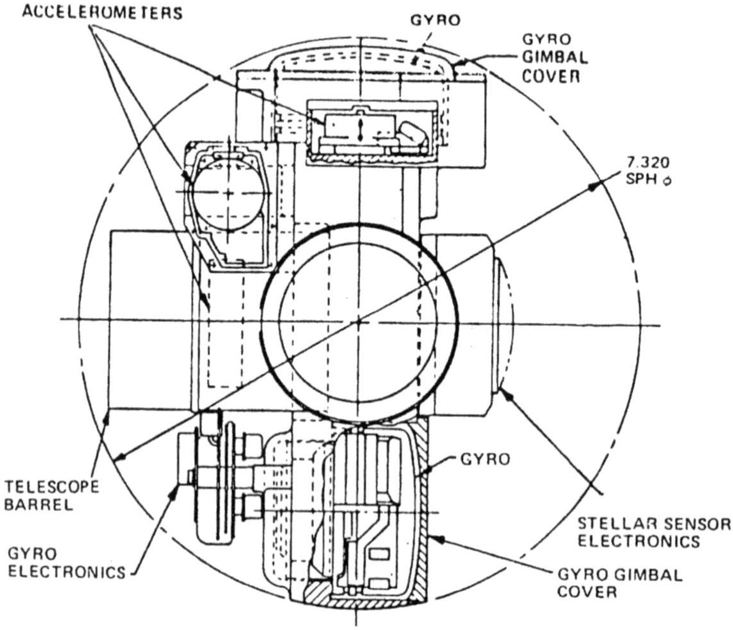

Fig. 5. Current 'paradigm' configuration of U.S. steller-inertial guidance system for ballistic missiles (Source: Stephen F. Rounds and George Marmar, 'Stellar-inertial capabilities for advanced ICBM', American Institute of Aeronautics and Astronautics, Guidance and Control Conference, 1983, Pibl. nr. 83-2297).

and areas where MMRBM was initially intended for deployment" (56). The development during the early 1960s of the idea of MIRV — Multiple, Independently Targetable Reentry Vehicles — however, made it possible to take the star fix *after* the boost phase, while the "bus" (postboost control vehicle) was "coasting" in space prior to releasing the reentry vehicles.

The Soviet Union and Stellar-Inertial Guidance

By the mid-to late 1960s, then, there was a growing consensus in the United States as to *how to design* a stellar-inertial system. As we shall shortly see, the desirability of such a system, and the need for it, remained in contention — with the result that it was not till 1979 that stellar-inertial guidance was deployed in an operational American missile.

These latter points do not appear to have been as contentious in the Soviet Union. But what is most interesting from our present point of view is not that this led the Soviet Union to be in advance of the United States in deploying stellar-inertial guidance in ballistic missiles, but that the form this guidance took is different. The theoretical import is clear. Once stabilization of a technology has taken place, it immediately seems that the successful design is the "only possible" one, the "best" one, the "most efficient" one. That is, indeed, in large part simply what stabilization *is*. To borrow an analogy from Harry Collins, a stabilized technology is like a ship in a bottle "with the glue dried and the strings cut. A ship *within* a bottle is a natural object in this world, and because there is no way to reverse the process, it is not easy to accept that the ship was ever just a bundle of sticks" (57). That in the Soviet Union stellar-inertial guidance has stabilized too, but *differently*, helps us remember that the American "paradigm design" is the *outcome of a process of negotiation*, not the simple product of "nature" or "technical efficency."

The circumstances in which stellar-inertial guidance came into operational use in the Soviet Union were ones that were also growing in importance in the United States. During the 1960s, both countries placed increasing reliance on submarine-launched ballistic missiles. These have a problem of initial conditions even worse than that of mobile land-based missiles (which at least can be known to be at rest with respect to the earth prior to firing). In both countries, early submarine-launched ballistic missiles were guided by a pure inertial system, with initial conditions being supplied by a separate inertial navigator — a Ships Inertial Navigation System, or SINS — in the submarine.

This approach was reckoned sufficient when the ranges of submarine-launched missiles were relatively short, as they were in the first part of the 1960s, especially given that great accuracy was not sought from such missiles (as we shall shortly see). But during the 1960s, NATO antisubmarine warfare capacities became more formidable, and the survival of Soviet ballistic-missile submarines patrolling close to the American coast was seen as potentially dubious. It would seem that to Soviet leaders a much longer-range missile began to appear attractive, because that would permit nuclear-missile submarines to patrol in ocean areas close to the Soviet Union.

But increased missile range placed increased strain on the combination of SINS and pure inertial missile guidance, if accuracy at the longer ranges were not to deteriorate. It seems to have been this consideration that gave stellar-inertial guidance in the Soviet Union its opportunity.

In 1969, the Soviet Union began to test a new submarine-launched missile with a much longer range (almost 8,000 kilometers) than its predecessors. Western intelligence analysts christened it the SS-N-8. Right from the beginning, at least one analyst was led to suspect that its guidance was stellar-inertial — the first test flight was at night. Analysis of intercepted telemetry soon confirmed the hypothesis. Western analysts felt sure they had detected telemetry recording corrections to the guidance system from a star sensor; interception of telemetry from the on-board computer even made it possible to deduce what particular corrections were being made. The follow-on SS-N-18, flight-tested in the mid-1970s, was agreed to be stellar-inertial (58).

Soviet systems have not, however, evolved in the direction of Unistar. In SS-N-18 guidance, multiple star sightings are taken (59), and reliance on an a priori statistical model is avoided, with gyro drift being deduced from repeated sightings in the manner outlined above. Since the Soviet designers could have read about Unistar in open Western literature at least as early as 1971 (and the first SS-N-18 flight test, for example, was not till 1976), it seems that ignorance cannot be the reason. *Why* Soviet designers have not accepted the Unistar argument is of course unknown. Presumably they have evaluated some of the above considerations differently. Perhaps difficulties with on-board computer technology have led them to feel that computational ease is worth purchasing at the cost of greater mechanical complexity. Perhaps their confidence in a priori statistical models is less. Perhaps, having chosen a design that they feel works adequately, they see no reason to change it. Perhaps, of course, issues of institutional power in the Soviet bureau responsible for submarine-launched missile guidance systems are involved. Whatever the explanation, their choice of a different paradigm design is of considerable interest.

Stellar-Inertial Guidance and the Fleet Ballistic Missile Program

By the late 1960s, then, the integral star sensor, taking a sight on one star once only, had become the dominant form of American ballistic-missile stellar-inertial guidance design. To that extent, "stabilization" had occurred. But this in no way guaranteed that the system would move from a paper design plus a few prototypes to large-scale production and deployment in actual operational missiles.

Air Force missiles offered little hope. With the cancellation of Skybolt, of the MMRBM, and of a plan to make Minuteman mobile, silo-basing was clearly going to remain for some time the norm for Air Force ballistic missiles. Arguments *could* be made that silo-based missiles could profit from stellar-inertial guidance. "Rapid reaction capacity" could be claimed, because of "the ability of the system to correct for initial azimuth error." So could "relatively low cost," because from this ability followed the capacity to fire rapidly without keeping the missile fully operational at all times. Further, "inertial sensor accuracy requirements for stellar-inertial systems are substantially relaxed as compared to all-inertial systems." But these arguments — if they were actually put forward in the period in question (60) — would have come up against what appears to have been a solid consensus that a star sensor was a wholly unnecessary addition to a silo-based missile.

As the Army remained devoid of long-range missiles, this left only the Navy. This was an attractive option. As explained above, submarine-launched ballistic missiles face an even more difficult initial conditions problem than the MMRBM. Further, unlike the MMRBM, the Fleet Ballistic Missile program was well established and insulated from immediate political pressures. A follow-on from Polaris, the Poseidon C3 missile, was already being planned; and by the later 1960s yet another system, ULMS, the Undersea Long-Range Missile System, was gathering momentum — this was the system that would eventually become Trident (see table 1).

On the other hand, the Fleet Ballistic Missile program already had a solution to the initial-conditions problem. The combination of a Ships Inertial Navigation System plus a pure inertial missile guidance system seemed wholly satisfactory to those most centrally involved. The Polaris program had developed an enviable reputation for technical, managerial, and political success. The accuracy of Polaris had exceeded expectations and seemed quite adequate for its primary "assured destruction" mission.

No wonder, then, that the developers of stellar-initial guidance at Kearfott should seek a home for their system within the Fleet Ballistic Missile System (61). But no wonder, too, that this was to be neither a quick nor an easy task.

The full story is complex, but it is clear that issues of several kinds were involved. They are worth examining, if only because they reveal the considerable heterogeneity of the factors that influence the nature of technical change in the area of nuclear weaponry. These factors from the "macro" (U.S. national strategy) to the "micro" (the organizational structure of the Special Projects Office of the U.S. Navy responsible for the Fleet Ballistic Missile program). Questions of desirability, feasibility, and necessity were also intertwined. Some officials continued to doubt whether stellar-inertial guidance would usefully enhance accuracy; others had no doubt that it would, but considered it a dangerous outcome; yet others felt that the same result could be better achieved by other means.

Stellar-Inertial Guidance and National Strategy

Perhaps the simplest level is the most "macro." There was little point in seeking to enhance the accuracy of the Fleet Ballistic Missile beyond the level considered achievable with a pure inertial Poseidon unless one was seeking the capability to destroy Soviet hard targets. Given that stellar-inertial guidance would usefully enhance accuracy (which does not appear to have been doubted at this macro level), was this desirable?

A significant force within Congress, led by Senator Edward Brooke of Massachusetts and organized by his energetic aide, Alton Frye, had serious doubts about the wisdom of the United States, developing a large-scale hard-target kill capability: might that not simply destabilize the strategic relationship between the United States and the U.S.S.R. by putting the Soviet Union in a "use 'em or lose 'em" situation in regard to at least the major part of its deterrent, the land-based missile force? When proposals were formulated to develop a stellar-inertial "Mark 4" guidance system for Poseidon, as an alternative to the pure inertial "Mark 3," the climate within Congress was thus potentially hostile. Brooke also enjoyed considerable access to the new Nixon administration. A moderate, black Republican, he occupied a key political role, which he used, in one instance, to extract a statement from Richard Nixon (in a December 1969 letter) that "there is no current U.S. program to develop a so-called hard-target MIRV capability" (62).

Importantly, Brooke's doubts were also represented within the Special Projects Office itself. In a situation where there was unlikely to be a clear political directive to develop hard-target kill capability, the attitudes of those in charge of the program were of some significance. Not only were they privately skeptical, but at least one important Department of Defense decision-maker, himself inclined towards a belief in the desirability of hard-target capability, recalls being successfully persuaded, by a leading Navy officer responsible for the program, that hard-target capability in the Fleet Ballistic Missile force was potentially destabilizing (63).

At this level of national strategy, however, a further consideration intervened to interrupt what might have become at least a temporary consensus that stellar-inertial guidance was undesirable. The counterpart to doubts about hard-target capability for the Fleet Ballistic Missile was a deep conviction that it must be a *sure* deterrent: that no conceivable Soviet action should be able to protect the Soviet Union from retaliatory destruction once it had struck the United States. The worry existed that a missile force with only the range of Polaris or a full-payload Poseidon — 2,500 nautical miles, or around 4,500 kilometers (64) — had to come dangerously close to the Eurasian landmass in order to fire, and might be exposed to attrition from Soviet antisubmarine warfare. A longer-range missile — perhaps 4,000 nautical miles, or maybe even 6,000 nautical miles — would have a great deal more sea-room within which to patrol, and would therefore be much safer.

But to maintain the same accuracy at greater range required both a better Ships Inertial Navigation System (since a given heading error, say, would translate into a proportionately larger target miss) and a better missile guidance system. So, paradoxically, by the late 1960s the "assured destruction" view of the Fleet Ballistic Missile System, as well as the "hard target" view, was generating a demand for enhanced guidance. This, then, provided the "macro" opportunity for stellar-inertial guidance to be seen as desirable by a wider constituency than simply those who welcomed hard-target kill capability.

National-strategic considerations therefore played a major part in shaping the context in which the decisions about stellar-inertial guidance were taken. Doubts about the desirability of hard-target kill were a factor in the process that led to the cancellation of Mk 4 stellar-inertial guidance option for Poseidon, thus leaving the Poseidon force equipped solely with Mk 3 pure inertial systems (see table 1). That the Trident I (C4) was designed as a longer-range missile but with the same final accuracy specification as Poseidon (65) meant that the strategic factors that had led stellar-inertial guidance to be controversial in the case of Poseidon were much less evident in this case; the decision to adopt stellar-inertial guidance for the C4's Mk 5 guidance system was thus, to my knowledge, one that was not contested by the opponents of hard-target kill capability. When the decisions were made about the Mk 6 guidance system for the Trident II (D5) there was no doubt that this missile was being straightforwardly designed for hard-target kill capability. But by then, this was politically much more acceptable than it had been a decade earlier (66). By then, too, the star sensor was an accepted part of the Fleet Ballistic Missile System technology, rather than an innovation that might attract political attack.

Stellar-Inertial Guidance and Bureaucratic Politics

These national-strategic considerations cannot, however, be wholly disentangled from ones that were political in a more mundane sense. The leadership of the Fleet Ballistic Missile program knew well that the health of the program was not simply a matter of how it fitted with national strategy: the wider reputation of the program, its relationship to the concerns of the rest of the Navy, and the potential for a damaging outbreak of interservice rivalry were all questions with which they had to deal. These matters of what is often called "bureaucratic politics" form the second type of issue that had a bearing upon the desirability of the star sensor.

Put simply, there were two different approaches to protecting and enhancing the program that could have been adopted. The first was that actually followed by the program's leadership, and it became a distinctive organizational and technical style. Its origins are well described by Harvey Sapolsky (67). It involved caution and moderation – an avoidance of taking on technical tasks, however attractive to political and military leaders, where there was believed to be a significant risk of failure. Overt conflict was also to be avoided, and potential opponents co-opted rather than defeated. What was distinctive about the Fleet Ballistic Missile System was to be emphasized. No attempt was to be made to enter into direct technical rivalry with Air Force programs; above all, there was no wish to seek to compete with them in hard-target kill capability. The autonomy of those running the program was to be preserved; those outside it who sought to impose their will had a difficult time if what they wanted ran counter to the judgment of the program's leadership, even if they were in higher formal positions in the hierarchy of the Navy or Washington.

The second approach was most strongly represented in one such higher location – the office of the Chief of Naval Operations. Here, particularly in the "Great Circle" group of planners, there was a desire to move away from the assured-destruction approach to nuclear strategy that was deeply evident in the Fleet Ballistic Missile program. Nor was this group daunted by the likelihood that a wholehearted attempt to pursue hard-target kill capability in Navy missiles would spark direct rivalry with the Air Force – they may indeed have relished this prospect.

The proponents of this second approach did seek to shape certain technical aspects of the Fleet Ballistic Missile program. They were, for example, advocates of equipping Poseidon with a much larger warhead (the Mk 17) than the Mk 3 that the Special Projects Office wanted (and that was, in the end, used). They were in favor of stellar-inertial guidance for Poseidon, "because it could make sea-launched systems as accurate as land-based missiles and, despite the loss of the Mark 17, would still remove the Air Force's unique claim on the counterforce mission" (68). Symptomatically, however, they lost eventually on this issue too.

The continuing dominance of the first approach made for an initially difficult climate for stellar-inertial guidance. Though it was a more mature "stabilized" technology than in the days of the MMRBM, the technical doubts about it, as we shall see were not wholly stilled: it could be seen as a "risk." Further, it would be a risk for a dubious payoff. Enhanced accuracy was a goal that the leadership of the Fleet Ballistic Missile program were not convinced was sufficiently central, either strategically or bureaucratically, that it should be pursued at, say, potential cost in terms of delay, unreliability, or exposing the program to controversy. Making stellar-inertial guidance a detachable option for Poseidon, which could be (and was) canceled without consequences for the rest of the program, and adopting it only later for the C4 once technical doubts were reduced and it was unlikely to stir up damaging controversy, made complete sense within the perspective of the first approach.

Economic interests also may have shaped attitudes to stellar-inertial guidance amongst the corporations involved in the program. It is self-evident that Kearfott developed an interest of this kind. During interviews, it was further suggested to me that Lockheed, which had and has overall systems responsibility for the Fleet Ballistic Missile System, came to support stellar-inertial guidance for a less obvious reason. As well as its overall responsibility, Lockheed was designing the small Mk 3 reentry vehicle for Poseidon, but not the large Mk 17; it thus had a considerable economic interest in the selection of the former. Stellar-inertial guidance came to count, in the argument about reentry vehicles, as an argument for the Mk 3, and against the Mk 17 — it was argued that it could give the Mk 3 the same capabilities as the Mk 17 on an all-inertial missile. This, it was suggested, led Lockheed to swing its weight behind the adoption of stellar-inertial guidance.

Will Stellar-Inertial Guidance Work? Is It Necessary?

Public strategic debate about stellar-inertial guidance — and much of the bureaucratic politics — took place on the assumption that the technology would in fact do what its proponents said it would do, and that it was a necessary means to achieve the goal of enhanced submarine-launched ballistic missile accuracy. It was thus debate about a technology whose characteristics were taken as known. But within the guidance community there was not unanimity that matters were as clear cut as all that.

The more influential form of doubt was the less radical. As outlined above, Charles Stark Draper's MIT Instrumentation Laboratory was deeply committed to the development of pure inertial technology. Improvement of the basic sensors was taken as the path to progress and enhanced accuracy. Of course, members of the laboratory knew that there were problems that inertial sensor improvement could not directly solve — but even in these a novel application of inertial technology could be sought, as in the ingenious development of an inertial "gravity gradiometer" to permit on-board correction of inadequate knowledge of the gravity field (69).

The attitude of Draper and his laboratory was of considerable weight in regard to the Fleet Ballistic Missile System, since the Instrumentation Laboratory has held — and continues to hold to this day — design responsibility for guidance, working directly for the Navy, not indirectly through a corporate prime contractor. Indeed, the criticism this major weapon-system responsibility attracted was a factor in the early 1970s divestment of the laboratory by MIT, and its independent establishment as the Charles Stark Draper Laboratory, Inc. (70).

There is some indication that initially there were those within the Instrumentation Laboratory who doubted that stellar-inertial guidance would work at all — who felt, that is, that the additional information from a star sensor would not increase accuracy (71). But that position, if it indeed existed, quickly disappeared from view. What replaced it was the different position that stellar-inertial guidance would work but was not necessary — that is, that the benefits it could bring could also be achieved from pure inertial technology. As one central figure in the laboratory's Navy work put it:

I must admit it [stellar-inertial guidance] wasn't pushed by us hard enough, it was pushed by Singer. I remember having tunnel vision myself talking with my boss at the time about a star-tracker: "Well, the system has been pretty good so far. We don't need to improve it with a star-tracker, necessarily" (72).

As a consequence, stellar-inertial guidance was not spontaneously adopted by the Instrumentation Laboratory as the path to follow. The proponents of stellar-inertial guidance first convinced crucial figures in the Special Projects Office, who then pressed consideration of it upon Draper's laboratory: "the Navy says . . . hey, Draper, Singer's got a good idea here, look into it" (73).

It would not be correct to portray the Instrumentation Laboratory as subsequently "resisting" stellar-inertial guidance. One interviewee claimed that the laboratory indeed put forward unsuccessfully its own stellar-inertial design, different from Kearfott's in involving an ultra-violet sensor mounted externally to the stable platform. Certainly, it was the laboratory's David Hoag who, in a 1970 Pugwash symposium, first highlighted the technology in its "stable" form in a public forum. His presentation endorsed it, albeit in a technically measured rather than zealous fashion (74). Some interviewees outside the Draper Laboratory do, nevertheless, continue to see enthusiasm there for stellar-inertial guidance as less than complete.

Perhaps there is a residue here of the situation of the mid-1960s, where the organization (Draper) centrally responsible for Fleet Ballistic Missile guidance was having a new technology from outside the program pressed upon it. Certainly, though, this does not now translate into real dispute about stellar-inertial guidance per se: stellar-inertial is now a stable feature of the program, no longer seriously questioned. If there remains a trace of the earlier situation in questions of technical design, it is at a more detailed level. I have no independent evidence of the assertion, but one interviewee suggested that the shift in star sensor from the Trident I's Vidicon tube to the Trident II's solid-state Charge Coupled Device was a shift from a technology in which the "incomer"

(Singer Kearfott) was adept, to a newer technology in which the established organization (Draper) could start on an equal footing and regain "territory" in the design process (75).

These possible shadings apart, there is no question now of the acceptance within the Fleet Ballistic Missile program of the idea that stellar-inertial guidance works. "Criticisms of a Unistar-stellar correction as a concept now differ little in credibility from doubts regarding ability to launch underwater," one officer told me; criticisms "have been discredited by now," he writes, and a critic would have "the credibility of a flat-earth proponent" (76).

Such doubts as remain in the guidance community outside the program are thus, in practice, unimportant. Nevertheless, for the analytical purposes of this paper they are worth examining. "Closure," in the sense used by the sociologist of science (77), has taken place — there is effective consensus that stellar-inertial guidance works. The arguments that can still be mustered against it, like the Soviet Union's "rejection" of Unistar, are of interest in revealing that this consensus is not self-evident to all rational, informed people.

These arguments were put to me privately by a guidance technologist who, though not part of the Fleet Ballistic Missile program, is acquainted with the technology used and with at least some of the test results (78). For him, the central issue is the adequacy of the statistical model necessitated by Unistar. The larger a given error source is expected a priori to be, then the more the requisite correction should be weighted towards an error of this kind. Both initial azimuth error and in-flight drift in azimuth of the relevant gyro in the missile's guidance system, say, will lead to the actual image of the chosen star being displayed in azimuth from the predicted image. A correction will then be applied in an attempt to reduce the cross-range miss of the target. But the best correction will depend on the relative extent to which initial error and gyro drift are contributing to the displacement: if initial azimuth error is expected to be much more significant than gyro drift, then the correction should be weighted towards initial error, and vice versa if drift error is expected to be greater. The proponents of the technology — from whom I have drawn the above example (79) — have confidence enough in the a priori statistical model not to regard this as a problem. This critic does not. He fears the model may be seriously wrong and — especially if this problem is combined with possible errors in the star map from which the predicted position of the chosen star is derived — the correction prompted by the star fix may increase, rather than reduce, target miss.

One might ask whether points like this were not finally settled empirically. Trident I (C4) was built with a star sensor, and by now it has undergone many carefully monitored flight tests. Those close to the heart of the program do indeed seem to regard these as having finally dispelled doubts about the feasibility of stellar-inertial guidance. The critic I am quoting, on the other hand, regarded the results of at least the first C4 test flights as being less than conclusive (80).

Again, issues familiar from recent studies of experiment in science reappear here in the testing of technology. What seems, on the face of it, to be a simple empirical issue turns out in practice to be complex. Stellar-inertial guidance is not assessed simply by observing the "fall of shot" (impact points) in missile tests (81). During the 1970s, especially with the Navy's Improved Accuracy Program, the sophistication and level of instrumentation of the tests of submarine-launched ballistic missiles grew considerably (82). At least six sources of data were employed. "Lonars"' a special version of the Loran-C radio navigation system, and "VPRS," a Velocity and Position Reference System using transponders on the ocean bed, were used to evaluate the performance of the Ships Inertial Navigation System and associated systems up to the time of launch. Shipboard data-recording monitored transfers of information and calculations performed in the submarine by the navigation, fire control, and guidance systems. "SATRACK," a tracking system using the experimental Global Positioning System satellites, measured the position and velocity of the missile in flight. In-flight telemetry relayed back what the missile guidance system "believed" about missile velocity and orientation, what the results of the star fix were, and what correction the guidance system then applied. Some reentry vehicles also contained telemetry packages that relayed information on what went on during and after their separation from the parent missile. Finally, acoustic sensors in the target areas were used to determine the points of splash-down into the ocean (83).

One aim of all this is to help produce and, crucially, to "verify" a mathematical model of the causes of system inaccuracy. Three points about this process may suggest what issues needed to be considered in the judgment as to whether flight-tests proved the efficacy of stellar-inertial guidance.

First, both stellar-inertial guidance itself, and the testing of stellar-inertial guidance, are, to use a philiospher's term, "theory-laden" in similar ways. Both involve use of the best possible model of system errors. This is not, of course, simply a circular procedure, because part of the point of the test procedure is to assess the model. The critic's point here is whether, even with the improved testing-range instrumentation currently available, the model can be tested adequately:

> There is inadequate time from launch to burnout for SATRACK or most of the other added instrumentation to really do a good job identifying an average gyro bias, much less how it changes in flight, so we go on guessing and giving statistics (84).

Second, it can be argued that assessing the model is not wholly "algorithmic" (85) – human judgment is involved. The case for this was put by Captain R. L. Topping of the Navy's Special Projects Office in a 1981 paper. Describing the last stage of model evaluations, he noted:

> Finally, the analyst interprets the results of [previous stages] and draws conclusions about model validity, potential model improvements, and the causes of system inaccuracy. For this step, there are no well defined methodologies. The analyst must rely heavily on experience, insight, judgment, and good communications with experienced colleagues in the Fleet Ballistic Missile Weapon System community (86).

Third, although I have no evidence that this has occurred in the Fleet Ballistic Missile program, in at least one publicly known case the error model for a U.S. missile system has been revised quite considerably. Testifying before the Armed Services Committee of the House of Representatives in 1976, Deputy Director of Defense Research and Engineering John B. Walsh described the recent history of the error model of the silo-based Minuteman force. He divided the error sources into four broad categories: guidance and control, which "refers to the guidance system"; gravity and geodesy, that is, "knowledge of the shape of the Earth, gravity of the Earth, and target location"; the process of separation of the reentry vehicles; and "deflection of the reentry vehicle as it enters the atmosphere." Walsh continued:

In 1971 the gravity and geodesy term decreased significantly and the accuracy was reduced. Notice at the same time our understanding of the guidance and control got better, which showed we had a greater guidance and control error — that in 1970 we were just wrong . . . .

In 1973 we took out some of the mistakes we had made in 1970 and the guidance and control error went down, and the total error went down.

But now look: as these large terms go down we began to wonder how to account for the errors we were observing, and concluded that the reentry dispersion was probably greater than we previously thought.

In 1973 the reentry error was carried as a large number . . . . Meanwhile the gravity and geodesy term goes down, which is supported by all the research the whole world is doing in the field of gravity . . . . But right now we carry no improvement in the reentry dispersion because it is unknown.

Maybe we will understand it better later. It is not clear we can make it go away (87).

Although it could be argued that the range instrumentation (especially SATRACK) available in Navy flight-tests is superior to that available in the Air Force flight-tests of the early 1970s, the critic quoted is not convinced that Navy error models are invulnerable to the kind of revision that might put the usefulness of stellar-inertial guidance in doubt: "I believe there would have been some improvement without the star trackers. How much improvement there is and what it is due to [are] not so clear" (88).

The Organizational Negotiation of Stellar-Inertial Guidance

Gradually, then, the feasibility, necessity, and desirability of stellar-inertial guidance within the U.S. Fleet Ballistic Missile program were established, and the remaining skepticism came to have little weight. All this did not, however, add up to completion of the process of stabilization of the technology — it had to be integrated in a practical way into the ongoing workings of a social and technical system.

Physically, the star sensor was of course to be placed inside the missile guidance system. Organizationally, that placed it firmly within the territory of SP-23, the branch of the Special Projects Office responsible for missile guidance and fire control, and of its guidance agency, the Draper Laboratory. The new technology did mean a new role within Navy missile guidance for the Kearfott Division of Singer. Potentially far more problematic, however, was the way the ramifications of the star sensor went beyond guidance. The whole point of the stellar-inertial system was to enable retrospective correction, from within guidance, of errors in the information coming from the Ships Inertial Navigation System. The latter was organizationally a separate domain, presided over by the navigation branch of Special Projects, SP-24, working with a quite different segment of the inertial navigation industry — notably, the Autonetics Division of Rockwell International in Anaheim, California.

The "technical characteristics" of stellar-inertial guidance, and its "organizational consequences," were therefore interconnected. To put it simply (we shall see shortly why it is *too* simple), if stellar-inertial guidance could satisfactorily correct for errors in initial conditions, then what the suppliers and designers of the submarine's navigation technology had to do would be reduced.

This could mean two things for "organizational interests" in stellar-inertial guidance and its characteristics. Stellar-inertial guidance could be seen as being against "guidance interests," because it complicated the task of guidance while simplifying that of navigation. Or it could be seen as being against "navigation interests" because it reduced the role and salience of navigation while increasing that of guidance. Both accounts were put forward to me in interview: the first, as explaining the reluctance on the part of SP-23 in the early 1960s to adopt stellar-inertial guidance; the second, as describing more recent attitudes.

My interview data do not permit me to assess either of these accounts. Both are "hearsay" assertions — from participants seeking to explain others' actions, or reporting on others' statements, rather than giving accounts of their own actions and statements. The following quotation, however, comes from an officer centrally involved in the negotiation over stellar-inertial guidance. With that technology,

> partitioning of the task became more difficult in that accuracy, unlike reliability, did not partition linearly. Even with exhaustive coordination, the accuracy performance of one branch could not be divorced from that of the others, and management visibility into

subsystem activity to a level beyond that desired by the branches became essential. Additionally, the process was affected by intra-organizational politics, budget realities and the capability and ambitions of the contractors (89).

Crucial to this process was, however, the realization that stellar-inertial guidance could not correct for *all* forms of error in initial conditions. That it could correct for errors in initial position and orientation was accepted, and this was indeed taken as meaning that increased missile accuracy could be obtained without concomitant increase in the accuracy of the position and the orientation information supplied by the Ships Inertial Navigation System. But it was also accepted that stellar-inertial guidance could *not* correct for errors in initial velocity, because these did not translate into an error in orientation of the guidance system with respect to the stars. The proponents of stellar-inertial guidance accept this as an inherent limitation on the technology, and deny that any responsible proponent ever argued that initial velocity error could be corrected. A central figure on the navigation side of Special Projects, on the other hand, stated:

[The star sensor] . . . alleviates performance problems on navigation. [But] some of them it doesn't address at all, like velocity. . . . It took us five years to convince [industry proponents of stellar-inertial guidance] that 1 + 1 = 2, and when it didn't address it, it didn't address it . . . (90).

This discrepancy of recollection may come about in part because "correcting errors in initial velocity" turns out not to be as simple a notion as it sounds (91). But whatever the past situation, that it is now accepted that stellar-inertial guidance cannot help with initial velocity errors means that navigation remains central. An excellent Ships Inertial Navigation System remains necessary; indeed, it may be necessary to supplement the velocity information derived from it by a precision Velocity Measurement Sonar, which "would be operated only moments before missile launching and would provide a very accurate initial velocity determination for the guidance initializations" (92).

Determining what the technical characteristics of stellar-inertial guidance were was thus part of the process of renegotiating organizational roles disrupted by its advent. Again, this appears to have been successfully achieved. Stellar-inertial guidance can now be taken as a technology whose characteristics are known and agreed upon and whose organizational consequences have thus been clarified and found acceptable by both guidance and navigation interests. A "stable technology," in other words – not one doomed, as others have been, by the overly great changes in organizational roles needed to make them viable (93).

Conclusion

What we have found here will not surprise the proponents of the "new" sociology of technology (94). I have tried to unravel some of the complexities of the process "of stabilizing" a new technology. We have found a continuous process of negotiation. That stellar-inertial guidance was feasible, desirable, and necessary was not self-evident – it had to be established, often in the face of skepticism, strategic objections, and competing technical approaches. The stellar-inertial navigation and guidance of aircraft and cruise missiles has become established in only a quite limited way in the face of these obstacles. Ballistic-missile stellar-inertial guidance has been more succesful; but it was finally adopted operationally, in the United States at least, a full decade after the proponents of the technology would have believed it to have been possible to do so.

This process of negotiation was *not* – as older approaches in the sociology of technology might imply – a matter of the negotiation, of the acceptance or rejection, of a technology whose optimal design and technical characteristics were unproblematic. Negotiating *acceptance* of the technology was simultaneously negotiating what it *was*. Only gradually did a stable form of ballistic-missile stellar-inertial guidance emerge in the United States; and meanwhile, a different form has emerged in the Soviet Union. Only gradually did effective consensus emerge that the technology did usefully enhance accuracy – consensus that even now is not quite universal. And there is at least some evidence that it was also only gradually that consensus emerged on more specific characteristics of the technology – that it could correct for errors in initial position and orientation, but not for those in initial velocity.

Further, the elements involved in this process of negotiation are genuinely heterogeneous (95). They include abstract argument (the Unistar principle); the organizational relationships of different branches of the Special Projects Office, and between different guidance contractors; the desirability or otherwise of the Navy's directly competing with the Air Force in the matter of accuracy; the wisdom of the United States' seeking the capacity to destroy Soviet missiles in their silos; and much else.

In the light of the issues raised in the introduction, however, a crucial question must be asked of this heterogeneity: is it layered? Does it divide up into "political" and "the technical," disputes about "ends" and disputes about "means"? Is there, for example, a "microsociology" of the technical community, concerning the negotiation of optimum design and technical characteristics, on the one hand, and a "macrosociology" of the political and strategic "context," concerning whether stellar-inertial guidance was desirable, on the other? (96)

In the case of stellar-inertial guidance and the Fleet Ballistic Missile System, it transpires that the answer is largely "yes." In marked contrast to the debate currently taking place over the Strategic Defense Initiative, stellar-inertial guidance scarcely became openly controversial, and such public "political" dispute about it as took place was insulated from "technical" dispute. Congressional opposition to stellar-inertial guidance was premised on the assumption that it *would* "work," that it would enhance accuracy. Whether that was actually so, whether stellar-inertial guidance was the best route to enhanced accuracy, how best to design a stellar-inertial system, what exactly the star fix corrected for — these matters were discussed, but in private. And the discussions took place largely within the technical community — they did not spill into the formal political arena; even the "bureaucratic politics" of stellar-inertial guidance was affected by them only in the sense that the aura of doubt surrounding the technology had to be dispelled before the Special Projects Office would adopt it. Nor do participants in technical dispute about stellar-inertial guidance seem generally to have deployed political arguments to support their case or undermine that of their opponents. Only at the most detailed level — that of the organizational structure of the Special Projects Office and its contractors — does it become difficult to distinguish technical from political (in this case, very "micro-political") matters.

Should this then be taken as indicating that the dichotomy between "political ends" and "technical means" discussed in the introduction is acceptable after all? Are social scientists justified in once again shutting the black box of technology, considering it as containing only the "uninteresting technical details" that are secondary to the "interesting" issues of bureaucratic, national, and international politics? Or, more sophisticatedly, ought we to conclude that at most we have a "micropolitics" or "microsociology" of the technical community, wherein technical matters are resolved, and which should be studied separately from the "macropolitics" of national policy and interservice rivalry?

I would argue *no*. First, it may be that the relative separation of technology and politics — or of micropolitics and macropolitics — to be found in the Fleet Ballistic Missile case is actually unusual. It is, for example, not to be found in the other case we have considered in detail, the MMRBM. That the MMRBM remained the "mobile political missile" made it difficult for it to become a stable technical object. Overly political and strategic considerations had, it appears, to be borne explicitly in mind in design decisions — they were not translated into stable "requirements," "technical specifications," or "design goals" as they generally have been in the Fleet Ballistice Missile System.

Second, even in the Fleet Ballistic Missile case the separation of technology and politics is not complete. One important "technical" proponent of stellar-inertial guidance, in interview, gave an interesting account of his role in bridging the separation. Through the then-president of the corporate division for whom he worked, he was introduced to members of the strategic community who had become convinced that the United States ought to give higher priority to missile accuracy than it currently (1966) was doing. He persuaded them that stellar-inertial guidance was what was needed: one of them, in turn, convinced his divisional president of its importance. The latter, a man with strong Washington connections, set up a meeting for the interviewee with the director of Defense Research and Engineering, Dr. John Foster, who shared strategic concern for missile accuracy. Foster then played a key role in pressing for the adoption of stellar-inertial guidance as an option for Poseidon.

As we have seen, that option was canceled. This interviewee then, as he put it, decided that the "only way to get this [stellar-inertial guidance] was to join the power structure," and he himself moved to a Department of Defense post, where he was involved in the negotiation of the specifications for the Trident I C4 that bore upon guidance — missile accuracy, missile range, and the maximum number of days of submerged cruise without access to external navigation aids. He ensured that these specifications were such that "only one system [stellar-inertial guidance] could do it" — that is, could fulfill the specifications. Simultaneously, he sought to avoid emphasizing counterforce capability in those specifications (despite his personal conviction that it was desirable), because of the likely opposition it would stir. Instead, the "necessity" of stellar-inertial guidance was established through specifications, notably the requirement for lengthy submerged cruise without access to external navigation aids, that would not offend the proponents of mutually assured destruction.

Here, then, the "means" — stellar-inertial guidance — was influencing the "ends" — the strategic characteristics of Trident, albeit in a way whose importance is hard to assess (97). Put another way, this engineer came to realize that to be successful he had to be a "heterogeneous engineer" (98), and deal simultaneously with the technical and the political.

What this points to is the third, and most important, reason for not returning to an uncritical means/ends, technical/political dichotomy. That the dichotomy does at least sometimes break down indicates that it is a contingent achievement, not an inherently necessary feature of technological change. It is crucial to be able to explain *why* in some cases technology and politics appear to be separate, and in some they do not. The dichotomy, in other words, is part of what we have to explain, not something that should be built into our studies a priori.

Thus the extent of separation of the technical and the political in the case of the Fleet Ballistic Missile program is both a cause and a measure of the success of that program. Though those involved would not express it precisely this way, the creation and maintenance of that separation has been for them a crucial, conscious goal. They have sought to avoid having their technical decisions shaped by politics. Their often-noted "technical conservatism" (99) — one reason for their caution in regard to stellar-inertial guidance — flows in large part from their consciousness that the black box of technology can be kept shut, can be kept separate from politics, only if it is seen as functioning smoothly. Public failure is the surest way to undermine the separation of the technical and the political (100).

It would indeed be tempting to suggest that the separation of the technical and the political is *in general* simply a cause and consequence of success. Certainly, the comparison between Trident and the highly politicized, and relatively unsuccessful, MX program points in that direction. But that conclusion might mistakenly identify one route to success, under particular circumstances, with the routes to success in general.

For example, in the early days of the Soviet missile program it is clear that both Stalin and later Khrushchev were involved, in a continuous, direct fashion quite unlike that of political leaders in the United States. Although the top Soviet designers were given substantial autonomy, Khrushchev claims to have come up personally with one major technical innovation — silo-basing — and to have pushed for it against the advice of the technical experts (101). Certainly, top political involvement in that early period seems to have been crucial to the success of advanced technological programs conducted under difficult circumstances (102).

We should therefore be cautious about the likely correlates of any particular pattern of relations between the technical and the political: it seems quite likely, for instance, that the Strategic Defense Initiative may presage a pattern quite different from the separation found in the case of the Fleet Ballistic Missile program or the substantial levels of political command found in the case of early Soviet missile programs.

Finally, however, we should note that even if the technical and the political are found to be separate it does not mean that the technical is an autonomous sphere shaped only by its "internal logic," or even by the "micropolitics" of the technical community alone. To avoid having their technical decisions determined by politics, the leadership of the Fleet Ballistic Missile program had to let them be shaped by the exigencies of maintaining the separation between the technical and the political. Thus, one of the reasons for not pressing as hard as possible for stellar-inertial guidance for Poseidon was the knowledge that it would be politically controversial to do so; one of the reasons for adopting it for the Trident C4 may have been the knowledge that it was unlikely to stir similar controversy. It may well be, therefore, that when we find technology not being shaped overtly by politics, it is actually being shaped by the need to maintain the contingent social achievement that is the separation of technology and politics (103).

There is, for example, no direct trace of politics in the Unistar argument that played such a key role determining the stable form of stellar-inertial design in the United States. But recall that a crucial advantage of a stellar-inertial system designed according to Unistar was its mechanical simplicity, and therefore, presumably, its ruggedness and reliability. Recall that those who had to be persuaded of the virtues of stellar-inertial guidance were aware that a condition of maintaining their autonomy, and freedom from political interference, was their reputation for the production of reliable systems, on time and without cost overruns. Recall, too, that they were conscious of the potential dangers — both strategic and in terms of making the program politically controversial — of giving extreme priority to accuracy, and were therefore unlikely to be overly concerned that the nonavailability of Unistar's "optimum star" might mean a 5% degradation in accuracy (104). Might we not then speculate that the stable form of stellar-inertial guidance in the United States was shaped, not directly by politics, but by the desire to keep technology and politics apart?

Acknowledgments

My grateful thanks are due to the Nuffield Foundation, who supported the research reported here, and to those involved in the story of stellar-inertial guidance who permitted me to interview them and who have corresponded with me about the matters discussed here. Responsibility for the views expressed here of course remains my own.

Notes

1. Strategic Systems Projects Office, *FBM Facts/Chronology — Polaris, Poseidon, Trident* (Washington, D.C.: Navy Department, 1982), pp. 4, 53; R. P. Berman and J. C. Baker, *Soviet Strategic Forces: Requirements and Responses* (Washing-

ton, D.C.: Brookings, 1982), pp. 106—107. See also B. Wright, *World Weapon Database. Vol. 1, Soviet Missiles* (Lexington, Mass.: Lexington Books, 1986).

2. R. D. Dalgleish and L. Schweikart, *Trident* (Carbondale and Edwardsville, Ill.: Southern Illinois University Press, 1984).
3. T. Greenwood, *Making the MIRV: A Study of Defense Decision Making* (Cambridge, Mass.: Ballinger, 1975), pp. 9, 65, 80, 136, 150. There is a remark on p. 65 that does not wholly fit this pattern of analysis: the suggestion that opposition to stellar-inertial guidance may in part have been "perhaps because it [the Navy's Special Projects Office] preferred to use only well-proven technology." I imply no criticism of Greenwood's useful analysis, which is drawn on below and forms an important part of what remains, to my mind, the best single case-study of strategic weapons development.

   Stellar-inertial guidance does of course also find occasional mention in literature concerned directly with nuclear-weapons policy issues. An early piece that grasped its significance was D. Ball, "The Counterforce Potential of American SLBM Systems," *Journal of Peace Research* **14** (1977), 23—40, esp. p. 32.
4. See T. J. Pinch and W.E. Bijker, "The Social Construction of Facts and Artefacts: Or How the Sociology of Science and the Sociology of Technology Might Benefit Each Other," *Social Studies of Science* **14** (1984), 424—429.
5. See B. Latour, *Science in Action* (Milton Keynes, Bucks.: Open University Press, 1987).
6. The centrality for a sociology of science and technology of the sense that technology "works" was noted by M. Mulkay, "Knowledge and Utility: Implications for the Sociology of Knowledge," *Social Studies of Science* **9** (1979), 63—80.
7. A historian of technology with enormous sensitivity to this is Thomas P. Hughes: see his *Networks of Power: Electrification in Western Society, 1880—1930* (Baltimore: Johns Hopkins University Press, 1983), and his fascinating earlier article, "Technological Momentum in History: Hydrogenation in Germany, 1898—1933," *Past and Present* **44** (1969), 106—132.
8. Latour, *op. cit.*, 1987, (5), p. 123.
9. See D. MacKenzie and J. Wajcman (eds.), *The Social Shaping of Technology* (Milton Keynes: Open University Press, 1985).
10. A classic example of the social significance of a design "detail" (the height of the overpass bridges on the Long Island parkways) will be found in L. Winner, "Do Artifacts Have Politics?" in *ibid.*, p. 28
11. Monte D. Wright, *Most Probable Position: A History of Aerial Navigation to 1941* (Lawrence, Kan.: University Press of Kansas, 1972).
12. David H. DeVorkin, "Electronics in Astronomy: Early Applications of the Photoelectric Cell and Photomultiplier for Studies of Point-Source Celestial Phenomena," *Proceedings of the Institute of Electrical and Electronics Engineers* **73** (1985), 1205—20.
13. See F. K. Mueller, *A History of Inertial Guidance* (Redstone Arsenal, Ala.: Army Ballistic Missile Agency, n.d.), now reprinted in *Journal of the British Interplanetary Society* **38** (1985), 180—192; H. Hellman, "The Development of Inertial Navigation," *Navigation* **9** (1962), 83—94; L. I. Tkachev, *Sistemy Inertsialnoe Orientirovki* (Moscow: MEI, 1973), English translation available as *Inertial*

Navigation Systems (Springfield, Va.: National Technical Information Service, 1976) (JPRS 67628); W. Wrigley, "The History of Inertial Navigation," *Journal of Navigation* **39** (1977), 61—68.

14. Not all actual inertial systems were configured in the way described in the text and in figure 1, but nevertheless something like this figure has been used for many years to introduce novices to the concept of inertial navigation. It is thus a "paradigm" in Kuhn's sense of "exemplar"; see T. S. Kuhn, *The Structure of Scientific Revolutions* 2nd ed. (Chicago: University of Chicago Press, 1970).
15. For some of the early skepticism and how it was overcome, see D. MacKenzie, "Missile Accuracy: A Case Study in the Social Processes of Technological Change," in W. E. Bijker, T. P. Hughes, and T. Pinch (eds.), *The Social Construction of Technological Systems: New Directions in the Sociology and History of Technology* (Cambridge, Mass.: MIT Press, 1987).
16. See E. Beard, *Developing the ICBM: A Study in Bureaucratic Politics* (New York: Columbia University Press, 1976), and R. L. Perry, *The Ballistic Missile Decisions* (Santa Monica: Rand, 1967) (P-3686).
17. C. S. Draper et al., "Fundamental Possibilities and Limitations of Navigation by Means of Inertial Space References," February 1947 (copy in library of the Charles Stark Draper Laboratory, Inc., Cambridge, Mass.).
18. "Reverse salient" is the term used by Thomas P. Hughes to describe a situation that inhibits the growth or development of a technological system; see, for example, his *Networks of Power* (7).
19. R. L. Doty and R. F. Nease, "Augmented Inertial Cruise Systems," in G. R. Pitman, Jr. (ed.), *Inertial Guidance* (New York: Wiley, 1962), p. 216.
20. R. B. Horsfall, "Stellar-Inertial Guidance Reduces Error," *Aviation Week and Space Technology*, 17 March 1958, pp. 73—79.
21. *Ibid.*, pp. 73—74, emphasis added. Draper's philosophy can best be seen in his *Importance of Research Directed toward the Development of Ultimate Performance for Inertial System Components* (Cambridge, Mass.: Charles Stark Draper Laboratory, Inc. 1975) (P-030). I am indebted to Rex Hardy, formerly chief pilot, Special Weapons Division, Northrop, for a lively and informative account of early flight-testing of stellar-inertial system (letter to author, 23 May 1985).
22. See Beard, *op. cit.*, 1976 (16), and Perry, *op. cit.*, 1967 (16).
23. P. J. Klass, "Titan Guidance Reliable, Accurate," *Aviation Week*, 12 May 1958, p. 93.
24. Horsfall, *op. cit.*, 1985 (20), p. 73.
25. R. L. Perry, *System Development Strategies: A Comparative Study of Doctrine, Technology and Organization in the USAF Ballistic and Cruise Missile Programs, 1950—1960* (Santa Monica, Calif.: RAND Corporation, 1966) (RM-4853-PR), p. 42. Other information in this paragraph is taken from K. P. Werrell, *The Evolution of the Cruise Missile* (Maxwell Air Force Base, Ala.: Air University Press, 1985), chap. 4.
26. R. L. Perry, "Commentary," in M. D. Wright and L. J. Paszek (eds.), *Science, Technology and Warfare: The Proceedings of the Third Military History Symposium, United States Air Force Academy* (Washington, D.C.: Office of Air Force History, Headquarters USAF and United States Air Force Academy, 1969), p. 119.

27. . . . *Reaching for Stars: NAS-26 Precision Astroinertial Navigation System* (Hawthorne, Calif.: Northrop Corporation, Electronics Division, n.d.). The system to which I am comparing it is the Honeywell electrostatic gyro "SPN-GEANS" system used in the B-52 mod G and H strategic bombers.
28. *Northrop: the Industrial Genealogy of Inertial Navigation & Guidance Technologies* (n.p., n.d., but c. 1983), p. 6.
29. *Ibid.*, p. 3.
30. "SR-71 to use Skybolt Guidance Sensor," *Aviation Week and Space Technology*, 23 November 1964, p. 22.
31. *Ibid.*
32. As Michel Callon puts it: "if . . . black boxes stay shut and stable . . . it is because actors use all their forces to keep them closed" ("Boîtes noires et opérations de traduction," *Économie et Humanisme* no. 262 [November—December 1981], 55 [my translation]).
33. Problems of this kind led to major financial losses by Sperry, the first corporation to enter the civilian inertial air-navigation market, which it was indeed forced to abandon. General Motors, the first successful entrant, also had to face several years of substantial losses before breaking even.
34. The clearest account of why a model of the gravitational field is needed is in D. C. Hoag, "Ballistic-Missile Guidance," in B. T. Feld et al. (eds.), *Impact of New Technologies on the Arms Race* (Cambridge, Mass.: MIT Press, 1971), pp. 22—24.
35. See J. M. Wuerth, "The Evolution of Minuteman Guidance and Control," *Navigation* **23** (1976), 64—75.
36. There was also a scheme, which never came to fruition, to make part of the Minuteman force mobile, and there were studies in the late 1950s of a Small Hardened Mobile ICBM.
37. According to *Aviation Week*'s report (see note 30), technology from the Skybolt stellar-inertial system was used in the SR-71 navigation system.
38. "USAF Wins Space Role Support; Overruled on Deterrent," *Aviation Week and Space Technology*, 12 March 1962, pp. 70—71.
39. P. J. Klass, "MMRBM to Use New Development Plan," *Aviation Week and Space Technology*, 2 April 1962, pp. 22—23.
40. "Ford Charges State Dept. Killed MMRBM," *Aviation Week and Space Technology*, 12 October 1964, p. 29. For the Multilateral force, see J. D. Steinbruner, *The Cybernetic Theory of Decision: New Dimensions of Political Analysis* (Princeton, N.J.: Princeton University Press, 1974).
41. L. Boda, "Mobile Mid-Range Missile Delayed Again," *Aviation Week and Space Technology* 26 March 1962, p. 16.
42. Philip J. Klass, "MMRBM Guidance Challenges Industry," *Aviation Week and Space Technology*, 23 April 1962, p. 89.
43. *Ibid.*
44. *Ibid.*
45. *Ibid.*, p. 90.
46. *Ibid.*, p. 89.
47. "Two Contracts survive MMRBM Cut," *Aviation Week and Space Technology*, 31 August 1964, p. 25.
48. On the analogies between scientific experiment and technological testing, see D.

MacKenzie, "From Kwajalein to Armageddon? Testing and the Social Construction of Missile Accuracy," forthcoming in D. Gooding, T. Pinch, and S. Schaffer (eds.), *The Uses of Experiment: Studies of Experiment in the Natural Sciences* (Cambridge: Cambridge University Press, 1988).

49. The success of the first STAFF test was announced in "Air Force Cites Good Test Data in First STAFF System Flight," *Aviation Week and Space Technology*, 19 April 1965, p. 36.
50. S. F. Rounds and G. Marmar, "Stellar-Inertial Guidance Capabilities for Advanced ICBM," American Institute of Aeronautics and Astronauts, Guidance and Control Conference, 1983 (83-2297), p. 849. I am grateful to Matt Bunn for a copy of this extremely useful paper.
51. *Ibid.*, pp. 853—854.
52. D. Hoag, *op. cit.*, 1971 (34), p. 93.
53. P. J. Klass, "MMRBM Guidance Techniques Described," *Aviation Week and Space Technology*, 19 November 1962, pp. 83—91; B. Miller, "Star Tracker Boosts MMRBM Accuracy," *Aviation Week and Space Technology*, 27 April 1964, pp. 99—103.
54. Miller, *op. cit.*, 1964 (53), p. 103.
55. Rounds and Marmar, *op. cit.*, 1983 (50), p. 849.
56. Miller, *op. cit.*, 1964 (53), pp. 99—100.
57. H. Collins, "The Seven Sexes: A Study in the Sociology of a Phenomenon, or the Replication of Experiments in Physics," *Sociology* **9** (1975), 205.
58. W. Beecher, "SIG: What the Arms Agreement Doesn't Cover," *Sea Power*, December 1972, 8—11. See also "Soviets Test New MIRV Warhead ICBMs," *Aviation Week and Space Technology*, 25 February 1974, p. 20. For the dates of SS-N-8 and SS-N-18 tests, see Berman and Baker, *op. cit.*, 1982 (1), 106—107.
59. R. T. Ackley ("The Wartime Role of Soviet SSNNs," *US Naval Institute Proceedings*, June 1978, pp. 34—42) notes that the SS-N-18's guidance system has the "capability for two celestial observations" (p. 37).
60. The arguments I am quoting here are actually drawn from a much later (1983) paper: Rounds and Marmar, *op. cit.*, 1983 (50), p. 849. This paper represents a major attempt to argue stellar-inertial guidance "back in" to Air Force missiles — an attempt that was nearly successful, in that stellar-inertial guidance was an option for the guidance of the Small ICBM. However, mobile basing of the Small ICBM is envisaged, as well as possibly some silo basing, and that clearly counted in the selection of stellar-inertial guidance as an option.
61. Those at AC Spark Plug had also done so earlier, in the Polaris days, but without success.
62. Quoted in A. Frye, *A Responsible Congress: The Politics of National Security* (New York: MacGraw Hill, 1975), pp. 69—70.
63. Interview.
64. Poseidon is in fact typically deployed with less than its full warhead loading, to permit greater ranges than this.
65. "The accuracy [of the C4] . . . was to be equivalent at 4000 nm to that of the Poseidon C3 at 2000 nm" (Levering Smith, Robert H. Wertheim, and Robert A. Duffy, "Innovative Engineering in the Trident Missile Development," *The Bridge (National Academy Engineering)* **10** [Summer 1980], 12). In fact, the Trident C4 program surpassed this accuracy goal.

66. Some of the reasons for this are discussed in D. MacKenzie and G. Spinardi, "The Shaping of Nuclear Weapon System Technology: U.S. Fleet Ballistic Missle Guidance and Navigation, I: From Polaris to Poseidon," *Social Studies of Science* **18** (1988), 418—63, "II: Going for Broke," forthcoming, *ibid.*, **18** (4) (November 1988).
67. H. M. Sapolsky, *The Polaris System Development: Bureaucratic and Programmatic Success in Government* (Cambridge, Mass.: Harvard University Press, 1972), esp. chap. 2.
68. Greenwood, op. cit., 1975 (3), p. 65. The Great Circle Group was established by Secretary of the Navy Paul Nitze in 1964; in 1967 it became the Office of Strategic Offensive and Defensive Systems, and in 1971 the Strategic Offensive and Defensive Division, OP-62 (*ibid.*, pp. 22—23).
69. For an account of this fascinating device, see Mark A. Gerber, "Gravity Gradiometry: Something New in Inertial Navigation," *Astronautics and Aeronautics*, May 1978, pp. 18—26.
70. For an account of this controversy, see D. Nelkin, *The University and Military Research: Moral Politics at MIT* (Ithaca, N.Y.: Cornell University Press, 1972).
71. Interviewees unconnected with the Instrumentation Laboratory asserted that an early 1960s report from there contained theoretical analysis leading to the conclusion that stellar-inertial guidance could not advance accuracy beyond pure inertial levels. I have not been able to trace any such document.
72. Interview.
73. *Ibid.*
74. Hoag, *op. cit.*, 1971 (34), pp. 91—94.
75. Interview. There is a picture of the D5's star sensor in *Aviation Week*, 13 February 1984, 174. For an interesting study of the introduction of CCDs in another context, see R. W. Smith and J. N. Tatarewicz, "Replacing a Telescope: The Large Space Telescope and CCD's," *Proceedings of the IEEE* **73** (1985), 1221—35.
76. Letter to author, 20 November 1986.
77. "Closure" refers to the ways in which scientific controversies — which, in principle, could be conducted indefinitely — actually come to an end. See, for example, H. M. Collins, "The Sociology of Scientific Knowledge: Studies of Contemporary Science," *Annual Review of Sociology* **9** (1983), 265—285.
78. These results are of course classified, and so his arguments were of necessity put to me at an abstract level, not illustrated by particular data.
79. Rounds and Marmar, *op. cit.*, 1983 (50).
80. I am unclear whether he has had access to the results of more recent flights.
81. One could imagine an experiment in which a sample of missiles was fired with the full stellar-inertial system operating and another sample fired with the star sensor "switched off"; but Trident C4 tests were not, as far as I am aware, conducted like this!
82. The origins and nature of the Improved Accuracy Program are discussed in MacKenzie and Spinardi, *op. cit.* (66), part 4.
83. R. L. Topping, "Submarine Launched Ballistic Missile Improved Accuracy," American Institute of Aeronautics and Astronautics, 1981 Annual Meeting and Technical Display (AIAA-81-0935), p. 3. Again, I am grateful to Matt Bunn for drawing my attention to this paper.

84. Letter to author, 9 March 1987.
85. For a general attack on the "algorithmic" view of knowledge, and some consequences of abandoning it, see H. M. Collins, *Changing Order: Replication and Induction in Scientific Practice* (London: SAGE, 1985).
86. Topping, *op. cit.*, 1981 (83), p. 5. By 1981 Special Projects had become the Strategic Systems Project Office, and subsequently it became the Strategic Systems Program Office. For simplicity's sake, I use "Special Projects Office" throughout the text.
87. House Armed Services Committee, *Hearings on Military Posture, Fiscal Year 1977: Part 5, Research and Development* (Washington, D.C.: U.S. Government Printing Office, 1976), p. 199.
88. Letter to author, 21 January 1987.
89. Letter to author, 20 November 1986.
90. Interview.
91. Thus the following hypothesis was suggested tentatively to me as an explanation:

> One of the facts that causes some confusion . . . is that velocity errors and the direction of the local vertical are coupled. That is, the velocity error that builds up in an inertial navigation system due to initial conditions and errors in gravity modeling will result in a Schuler frequency oscillation in position error. This has both an initial velocity component and also a tilt of the vertical. Velocity error is sometimes the generic term for this, since the instrumentation errors are integrated to give velocity errors before another integration results in a position, and since the contributions of the Schuler oscillation cause more on-target error due to velocity than their equivalent position error. The fact that the position error is related to the orientation leads some people to say that the star sighting corrects the initial orientation which was caused by the velocity error. That argument is probably logically okay, except that at the instant of launch, it is not the velocity itself that is being corrected, but rather what it has propagated into as a position error through the Schuler oscillation (letter to author, 9 March 1987).

"Schuler oscillation" refers to the characteristic 84-minute period of oscillation of errors in an inertial navigator on the earth's surface, 84 minutes being the period of a hypothetical pendulum on the surface of the earth with its bob at the earth's center, or of a hypothetical satellite orbiting the earth at surface level. It is named "Schuler" after Max Schuler, "Die Störung von Pendel- und Kreiselapparaten durch die Beschleunigung des Fahrzeuges," *Physikalische Zeitschrift* **24** (1923), 344—350.

92. Topping, *op. cit.*, 1981 (83), p. 7.
93. The classic study here is of course Elting Morison, *Men, Machines and Modern Times* (Cambridge, Mass.: MIT Press, 1966).
94. See Bijker, Hughes, and Pinch, *op. cit.*, 1987 (15).
95. See John Law, "Technology and Heterogenous Engineering: The Case of Portuguese Expansion," in *ibid.*, pp. 111—34.
96. D. MacKenzie, "'Micro' versus 'Macro' Sociologies of Science and Technology?", paper read to the British Sociological Association, 6 April 1987.

97. Certainly it would be quite misleading to suggest that this was the only factor influencing the specifications for Trident.
98. A term coined by John Law, *op. cit.*, 1987 (95).
99. See, e.g., Greenwood, *op. cit.*, 1975 (3), p. 34.
100. For further discussion of Special Projects' "style," see MacKenzie and Spinardi, *op. cit.* (66).
101. N. Khrushchev, *Khrushchev Remembers: The Last Testament* (London: Deutsch, 1974), pp. 48—49.
102. D. Holloway, "Innovation in the Defense Sector," in R. Amann and D. Cooper (eds.), *Industrial Innovation in the Soviet Union* (New Haven, Conn.: Yale University Press, 1982), pp. 276—367.
103. There is an analogy here with the separation of "economics" and "politics." It would be quite wrong to see politics as being determined by economics. However, both "political life" and "economic life" in our kind of society are shaped by the *separation* of politics and economics, and by attempts to shore up that separation, or maintain the illusion of it when it breaks down. See J. Holloway and S. Picciotto, "Capital, Crisis and the State," *Capital and Class*, no. 2 (Summer 1977), 76—101.
104. See Rounds and Marmar, *op. cit.*, 1983 (50), p. 854.

EUROPEAN STAR WARS: THE EMERGENCE OF SPACE TECHNOLOGY THROUGH THE INTERACTION OF MILITARY AND CIVILIAN INTEREST-GROUPS

JOHANNES WEYER

University of Bielefeld

Introduction

A fundamental shift for preparing the entry into a new, militarily relevant big technology is currently taking place in West Germany. This technology — its proponents maintain — can be compared in terms of its dimensions, costs, and social consequences with nuclear energy. I am referring to West Germany's plans for manned space flight, which have been outlined by a number of decisions during 1985 and 1986; they will eventually lead to a national space program and thus to a fundamental reorganization of research priorities. While it is still impossible to predict the outcome of these ongoing processes, we have the chance of observing the emergence of a new big technology out of a specific area of social, political, economic, and military interests; we can also observe the actions of those interest-groups, which begin to form a stable pattern of argumentation and legitimation, thus producing, step by step, a self-consolidating social structure with its own interests and its own dynamics.

The space lobby, however, doesn't start from zero; a system (in terms of social force and the power to determine research policy) to some extent comparable to the U.S. military-industrial complex has been established during the last thirty years. It possesses an inner coherence and institutional resistance. The development of this complex is one important point this paper will examine. German plans for

E. Mendelsohn, M. R. Smith and P. Weingart (eds.), Science, Technology and the Military, Volume XII, 1988, 243—288.

manned space flight can be regarded to some extent as a continuation of (nuclear) high-tech policy in another field — that is, as a sort of follow-up technology for the defense-nuclear-space complex (1).

In addition, there is a political continuity, for space technology is viewed by many space-lobbyists as a unique chance for West Germany to catch up with the military superpowers again by participation in the postnuclear arms race. Space plans entail a continuity and at the same time a new quality of West German foreign, defense, and research policy. The frequent publicly pronounced reflections about a "European space-power" (Helmut Ulke, in Stellungnahmen, p. 7) under West German "system-leadership" (DGAP 1986:49), with well-known and openly discussed military consequences (2), indicate a new determination of the military-industrial complex, no longer to restrict itself to the development of civilian technologies (which was the trademark of West German research policy up to now [see Haunschild 1986:61]), but to insist on the interchangeability of civilian and military technologies, thus opening a wide field of common activities for science, industry, and the military.

This paper will first take a look back at the history of German space flight from the 1930s up to 1987 (section I), thus providing background information to help us to understand the processes that brought about the recent space plans and the groups involved in the game. These groups and their specific interests and argumentation strategies will be examined in a second step (section II), which will also show the mechanism that brings these different groups together and ties them up into a self-consolidating new social structure. Section III will describe and analyze the political program of the space lobby and discuss the financial and social costs that will emerge from it. The manner in which the space lobby deals with these problems which emerge when it confronts the public and the legitimation strategies used in these public debates form a distinct part of the identity of the space lobby and can be analyzed as a sort of system-environment relation that defines the borders of the structure and provides it with legitimacy. Those processes are analyzed in section IV, which also will discuss the relationship between civilian and military space flight. The final section tries to integrate the argumentation lines of the previous sections and to elaborate on some aspects of a sociological approach that serves as a (mostly implicit) analytical framework.

I. A Short History of German Space Flight

In order to understand the present situation in West German space discussions we need to look back briefly at its history, concentrating on manned space flight. (For illustration, see figure 1.) Though the early pioneers in the 1920s and 1930s often dreamed of man on the moon, the first institutionalization of space activities took place under military sponsorship during the Nazi era. As early as 1932 the Heereswaffenamt (Army Ordnance Department) — "naturally" interested in ballistic missiles of any kind — started support for the Nebel-Braun Group, and it installed one of the first (the world's first?) "Big Science" laboratories in Peenemunde in 1937. On the other hand, the Luftwaffe (Air Force) — "naturally" more interested in any sort of plane-like object — not only supported the construction of the cruise missile called V-1, but starting in 1937 it also gave Eugen Saenger the chance to construct new rocket engines as well as to design his antipodal space bomber, an aircraft similar to the U.S. space shuttle. Just as in the case of the Manhattan Project, in 1942 the command was given to start the mass production of A-4 missiles parallel to the research work still to be done. And in 1944, when the first military version was operational, the A-4 was renamed "Vergeltungswaffe" V-2 (retaliatory weapon) (3).

In 1945 space science, along with military R&D, was abolished in Germany, but research and construction went on in the United States, where Wernher von Braun and his team further developed the A-4/V-2 and finally, many years later, constructed the Saturn rocket. The Saenger shuttle disappeared in the files for a while until its revival in 1964. The German aerospace researchers spent the ten years that were to pass before military, nuclear, aeronautics, and space research were permitted again, in different fields. Some went abroad — for example, to Argentina — to continue research (4), or they worked on joint projects — for example, with France (Büdeler 1978:79). The Dornier company switched over to textile-machine building until they were allowed to continue their aerospace work (Büdeler 1978:109). Ludwig

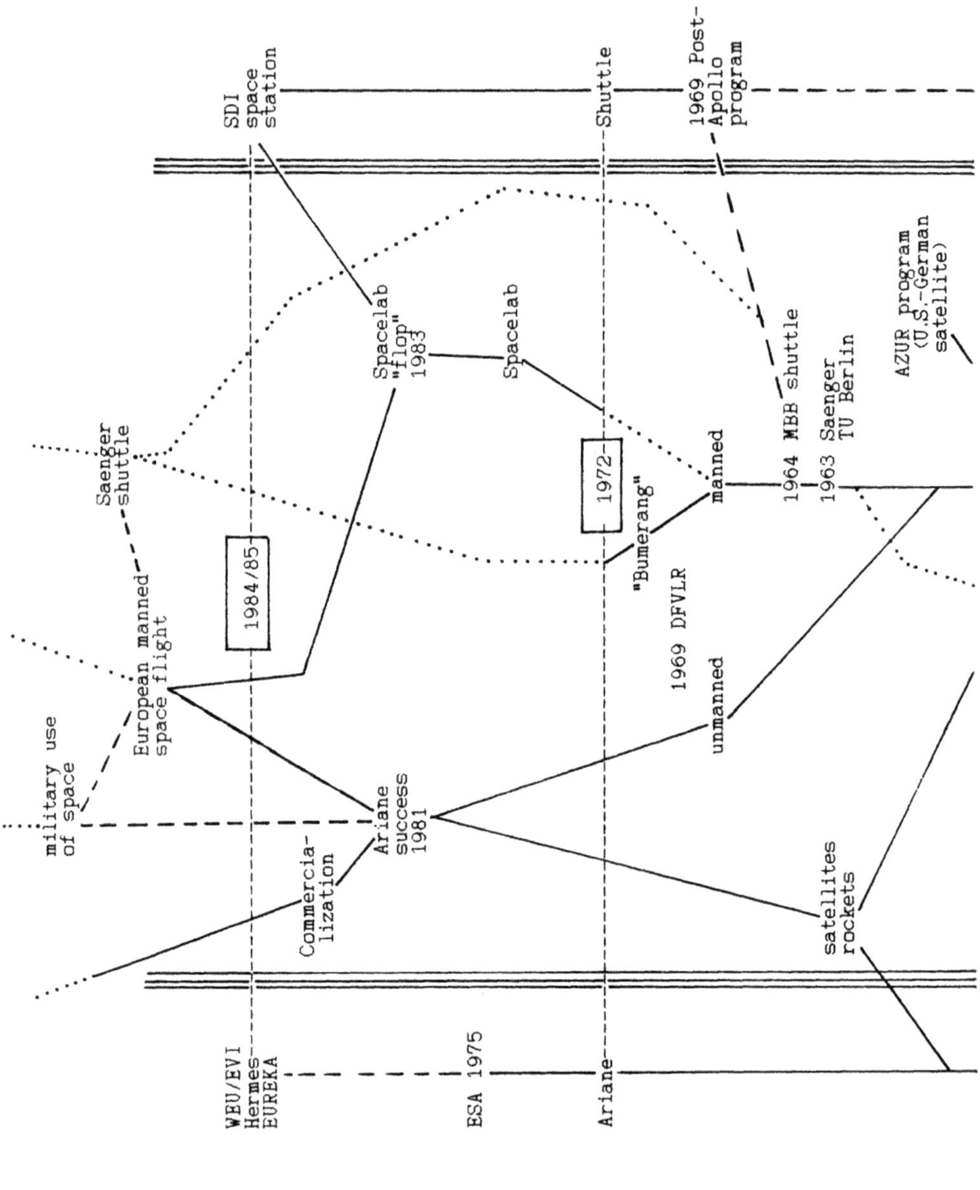
SDI space station
Shuttle
1969 Post-Apollo program
Spacelab "flop" 1983
Spacelab
1964 MBB shuttle
1963 Saenger TU Berlin
AZUR program (U.S.-German satellite)
Saenger shuttle
1972
manned
"Bumerang"
1969 DFVLR
1984/85
European manned space flight
unmanned
military use of space
Ariane success 1981
Commercia-lization
satellites rockets
WEU/EVI Hermes EUREKA
ESA 1975
Ariane

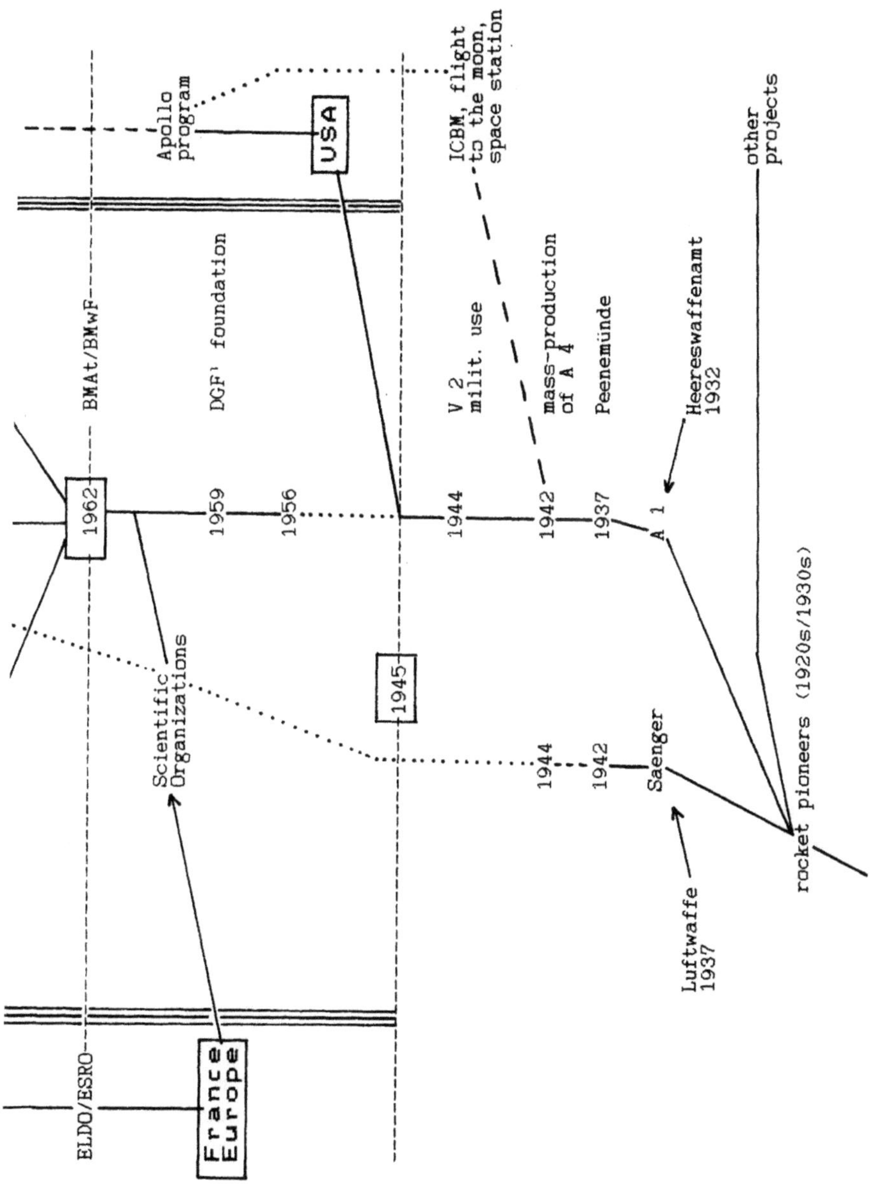

Fig. 1. History of German space flight (Source: Weyer, 1987). [1] Deutsche Gesellschaft für Flugwissenschaften (German Society for Aeronautic Sciences).

Bölkow, the famous aircraft designer, worked for ten years "out of his branch" (Büdeler 1982:71) but reentered aeronautics in 1954 (in the fields of sports planes and helicopters); in 1956, however, he switched back to his original profession and started to construct rocket engines again. This was supported by research contracts from the newly established Department of Defense and the Ministry of Education and Science (Büdeler 1982:74). The Deutsche Forschungsgemeinschaft (German Research Council) also started its support for space research during this period.

With this recognition of space research, including rocket research, and with the beginning of public funding, the era of illegality and of relabeling and sidestepping was definitely over. But another six years would have to pass before space research became an issue for the government — largely due to proposals of European space scientists to establish a European organization for space research, which would include West Germany. The Deutsche Forschungsgemeinschaft was asked by the government in 1960 to put forth proposals for German space activities. This led to German participation in the foundation of the two European space organizations ELDO (European Launcher Development Organization) and ESRO (European Space Research Organization) and to the institutionalization of a department of space research within the Ministry for Nuclear Questions in 1962, which, among other things, can be held responsible for the expansion of that ministry to the Ministry for Scientific Research in 1962 (see Stamm 1981:229—230). From 1967 a special space program has been worked out, usually covering a period of about four years.

The 1960s can be regarded as the phase of reentry into the field. German space policy was concentrated on catching up with the competition — building up a space infrastructure and exploring the niches in the field left to the late-comer. This was achieved when the first German satellite went into space and at the same time the research capacities were concentrated in the Deutsche Forschungs- und Versuchsanstalt für Luft- und Raumfahrt (DFVLR); the coincidence of both occurrences in 1969 was probably accidental. During this phase some very important decisions were made. One affected the pursuit of the more traditional space sciences (those looking up to the stars and exploring the nature of space) and simultaneously the applied space sciences, above all rocket engineering; this was a distinct reorientation of the fields of space science. Another decision was of similar relevance: To accept the invitation to participate in joint European endeavors, but at the same time to start a national space program and different forms of bilateral cooperation (which are naturally less controllable). From the very first, Germany tried to make its space policy independent by way of commitments in different contexts. As figure 1 shows, these decisions in the early 1960s make present politics more understandable.

The American and the European connections were both very prominent for the promotion of different lines of technological development and for providing the frame of reference for a gradual selection from the widespread range of alternatives generated by space scientists. There was a host of sometimes incredible proposals generated in this utopian stage, in which German scientists and aerospace companies could do hardly more than close the gap to competitors, while at the same time preparing ideas for the future. Some of those ideas fitted into the existing framework. For example, Messerschmitt-Bölkow-Blohm (MBB) began in 1961 to design a high-energetic rocket engine (a partly military, partly European-civilian project), which is still used in the U.S. space shuttle (and is well paid) (Büdeler 1982:74—75). Project Neptun — a gigantic rocket, designed to operate in the post-Saturn age — never left the drawing boards in Berlin (5); the idea of achieving a European autonomy by constructing such a vehicle was simply too futuristic. But the plans for the Saenger space shuttle — which was redesigned in 1964 by Junkers (today MBB), and submitted in 1965 as a proposal for German-American cooperation (Büdeler 1978:40; Büdeler 1968:218, 236; Büdeler 1979:463) — and the participation of the aerospace company ERNO (Entwicklungring Nord) in the preshuttle program by designing and testing the shuttle "Bumerang" in the late 1960s (6), indicate that some of the concepts (based on specific German research traditions) were not entirely unrealistic.

Though Germany owed a lot to the United States because of the generous transfer of know-how and the free disposal of launching facilities (see Keppler 1986:532—533, Büdeler 1968:239—246), it was not willing to become a subcontractor in the U.S. post-Apollo program. Simultaneously with the U.S. offer for participation in this program, the Europeans were forced to think about the future of their own space program, since the rocket Europe turned out to be a flop and the conditions under which U.S. launching facilities could be used and the style of behavior became increasingly unfavorable (see, e.g., Büdeler 1979:371).

Decisions taken in 1972 mark the beginning of another stage in German space policy. They were:

- to launch the Ariane rocket program (a French proposal);
- to reorganize European space policy (which resulted in the foundation of the European Space Agency [ESA] in 1975); and
- to contribute to the U.S. post-Apollo program (the shuttle program, also settled in 1972) with the Spacelab, thereby occupying a niche in the program (see Köhler 1976:78).

With Spacelab, plans for a European shuttle were dropped, but the entry to manned space flight was programed. Applied space sciences and space engineering were given a higher priority than before. At the same time, Germany strengthened its position within ESA, because Spacelab was constructed under German leadership (and more than half of the costs were paid by Germany) (see DFVLR 1984:169). The ESA program was given a specific shape, consisting of the French-dominated Ariane program, which led to a breakthrough in the field of launchers and a commercialization of the satellite market, and the German-dominated Spacelab program, leading to future scenarios of manned space flight and other, unforeseeable, applications. Spacelab flew in 1983, but the investments were lost — in that Spacelab now is the property of NASA, and Europe has no access to this field except by using the shuttle every two or three years (e.g., the D-1 mission). Ariane, on the other hand, flew in 1981 and established a solid cornerstone of the so-called European autonomy in space.

The 1970s can be regarded as a stage in which Germany caught up with her competitors by means of different forms of cooperation and slowly started to stake her own claims in the field, while exercising influence on ESA politics in a way that led to European-American cooperation and to manned space flight. German scientists gained know-how in nearly every field and achieved a top position in some areas (satellite propulsion, space physiology, some fields of remote sensing). Not least important, the German organizational framework was reconstructed as concentration processes in industry left only two big high-tech aeronautics-aerospace-military trusts: Messerschmitt-Bölkow-Blohm and Dornier (nowadays, Daimler-MTU-AEG-Dornier). The forfeited investments in Spacelab are regarded by members of the space lobby as worthwhile because of that program's catching-up function (7).

The next stage, starting in 1984, shows a new style: that of a self-conscious German space policy, which is becoming more and more part of an integral political concept. This is now outlined by some think-tanks, which openly discuss plans for a future German superpower (based on space weapons) and sometimes do not hide their nationalistic, militaristic, and imperialistic attitudes toward the "high frontier" (8). Even if official statements of the government are much more moderate, the concepts of the space lobby are nevertheless compatible with the new politics carried out under the label "autonomy" — that is, Germany is no longer content with its role as a co-player but claims leadership in European space politics. This trend even allows discussion of the military applications of space technology now available (or under construction) in the German or the ESA arsenal. One essential factor of these plans is the pursuit of an independent European access to manned space flight within the next decade — even if a tactical cooperation with the U.S. space-station program is still in the game (9).

The main steps were taken in 1984 when President Reagan invited the Europeans to participate in the project of a space station and thus exerted decisive pressure on them. In 1984, too, the Western European Union (WEU), a defense alliance, discussed plans for a space-based European Defense Initiative, which would use such ESA technologies as, for example, the weather satellite Meteosat. And in 1985 ESA set up a new program containing the new rocket Ariane V and the European contribution to the U.S. space station called Columbus.

This short presentation of the history of German space flight can be divided, as we have seen, into three stages:

(i) 1961—1972: Reentry into the field, construction of the infrastructure for space research.
(ii) 1972—1984: Catching up with competitors, different forms of cooperation.
(iii) since 1984: Claim to a leading role, entry to manned space flight, trend toward a militarization of space flight.

This history also shows that there have always been different lines of technological development that were selected by political decisions. For example, the German shuttle Saenger has been in the game as one potential option for the last fifty years, and yet it has never been constructed (see figure 1). The development decisions were affected by the political and economical frame of reference, policy and research traditions, and different sets of interests.

I will attempt here to analyze the various interest-groups and the way in which they formed, step by step, an interest-alliance to realize the technology of manned space flight. Only an analysis of this kind can tell us why the development, presented above, proceeded in such a straightforward manner and how the space lobby could gain so much strength that is is now in a position to define the future of the nation as dependent on space flight (in much the same way as the nuclear lobby, in recent decades).

II. Constructing Coherence: Interest-Groups in the Space Arena

We can clearly identify four social groups in the space scene, some centers of reference in the environment, and a type of irreversible history characterized — as shown above — by a number of decisions made in the past, a specific tradition, and political inertia (for illustration, see figure 2). My main thesis is: The mechanism producing a novel social structure with a certain type of internal coherence can be defined by processes of adjustment of interests and transformation of arguments. Thus, different actors acting in formerly rather independent fields strengthen their own position inside those fields because they

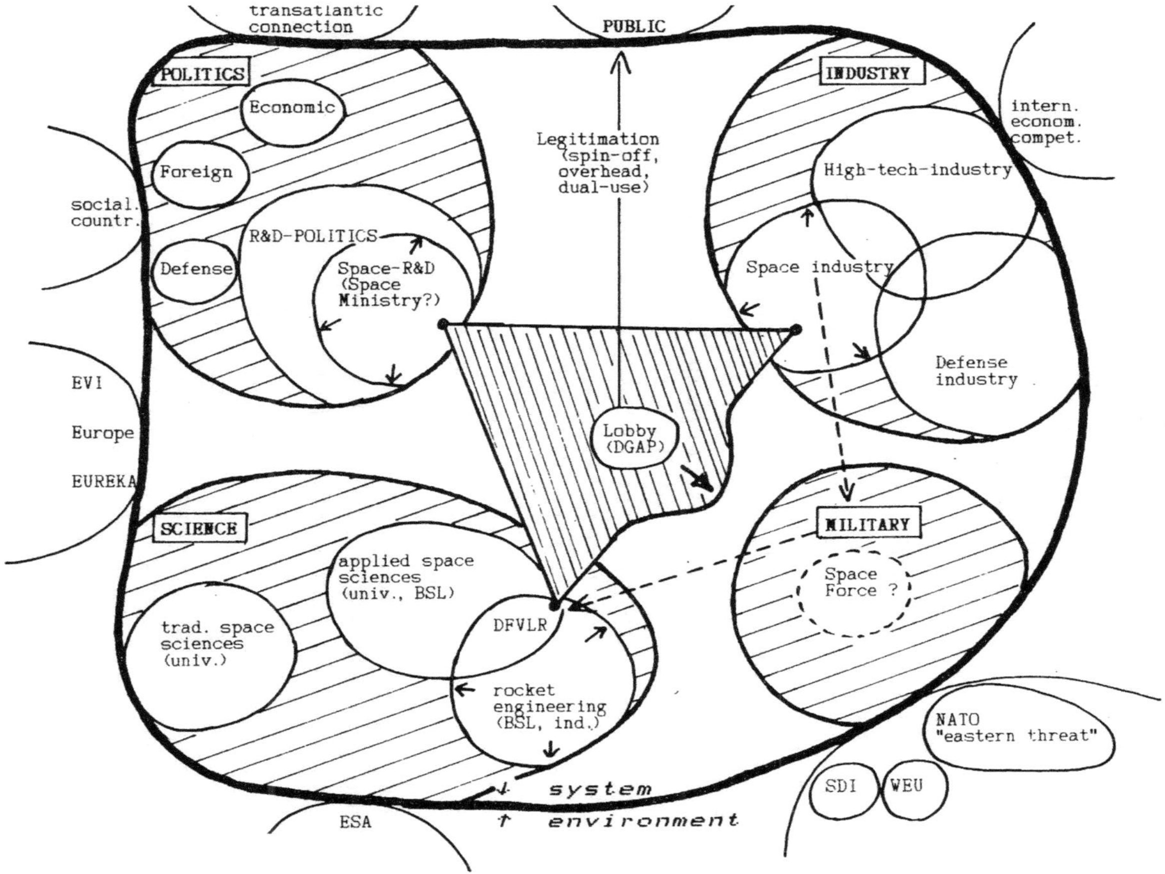

Fig. 2. The structure of the space lobby (Source: Weyer, 1987). BSL — Big Science Laboratories.

become able to import credibility from outside. These processes of mutual reinforcement, which build up hybrid or intermediate structures, subsequently lead to a restructuring of those fields in a way that can be described as a shift in the balance of power (10).

Science

Besides the traditional forms of space science, like astronomy, rocket engineering was established as a systematic purpose in the early twentieth century, and was pushed forward when the German military perceived the implications of that technology for revolutionizing warfare. The technicalization of traditional space sciences by the application of modern rockets (11) allowed the scientists in a first phase to complete their research and to solve puzzles that were unsolvable by terrestrial technology — in a way, to pursue their research paradigm and to achieve some revolutionary discoveries (see Keppler 1986:532). But the constant technicalization of space projects, which was additionally reinforced by manned space flight, on the one hand, and a sort of dynamics of space flight on the other, led to a remarkable paradigm-shift in space sciences due to the fact that, if transported into space, instruments and telescopes could be turned around to look back at the earth. Earth remote sensing came into being as a field of (largely military) interest. The main shift occurred when space exploration was gradually replaced by research in space and research on problems of manned space flight (see Feuerbacher 1984:53, and Köhler 1976). This step was carried out in Germany by the Spacelab program.

The scientists concerned with the more traditional fields, which earlier had profited from unmanned space flight, now complain about space politics and the neglect of their concerns. They have only a few arguments left (mostly the general cultural benefits of science and the technological spin-off from utopian scientific programs [see Keppler 1986]). They have no allies in other fields, and the gradual displacement of scientific organizations from the decision-making process is obvious. In 1960 the Deutsche Forschungsgemeinschaft still outlined the German space program (see Gambke, Kerscher, and Kertz 1961); but with the constantly increasing dominance of the space industry and applied science and engineering, the traditional scientific organizations (representing the academic branches) have lost their influence. In 1984 the main organizations wrote a letter to the research minister protesting the funding of Columbus out of the research budget (12), which was combined with the cancellation of some basic research projects (e.g., the German infrared telescope GIRL) — but they could not exercise any influence. The classical space sciences are reduced to a marginal position in the field, and former outsiders, who are linked up in a network with other actors and who have the "better" arguments, now have the power to define the course of action.

Those former outsiders are: (i) the applied space sciences, which emerged as a by-product of the new space technologies (partly as quasi-civilian copies of military fields of interest), and (ii) other subfields in the area of rocket engineering, which have been systematically built up during the 1960s (see Büdeler 1968:233—234). Both types of research are institutionalized in the German Grossforschungseinrichtungen (Big Science Laboratories), the most prominent of which is the Deutsche Forschungs- und Versuchsanstalt für Luft- und Raumfahrt (DFVLR); this is not only a large research center, but the parent organization of nearly all space research, with executive and advisory functions as well. The DFVLR acts as a brain-trust for the government, and it is connected in different ways with industry and the military. It receives one-fourth of its federal funding from the Defense Ministry (BMVg) and the remaining three-fourths from the Ministry of Research and Technology (BMFT) (in some fields, like microwave remote sensing, the BMVg pays even more). Only in space flight is the military totally absent, a fact that will be discussed later (see DFVLR 1986:9/2).

In a "Strategy Study of Space Flight," released in 1984 as an expertise for the BMFT preparing the European decisions for 1985, the DFVLR recommended German participation in manned space flight as the best of a large range of alternatives. In the study those alternatives are seriously discussed, the costs are carefully calculated, and the conclusions are rather cautious. For example, it is estimated that a final decision about the worthiness of manned space flight cannot be passed before the end of the century; the study also recommends viewing participation in the U.S. space station as "an experimental bridging phase" (DFVLR 1984:xii), which doesn't require the building up of a separate infrastructure. A comparison of these considerations with the argumentation presented by the chairman of the DFVLR, Prof. Hermann L. Jordan, during the parliamentary hearing in 1985, shows distinct variations. Here he characterized manned space flight as "indispensable" (Anhörung, p. 153) and the construction of a European hardware as "a necessary step" (*ibid.*). A "simple alternative" (p. 157), he continued, doesn't exist; when asked for the costs, he answered: "It costs money" (p. 233). In his written opinion he even demanded a "better connection of civilian and military projects" (Stellungnahmen, p. 7) in the German space program and a participation of the BMVg "in using surveillance satellites" (p. 4). During the hearing it was apparently not opportune to push that argument too strongly (Anhörung, p. 36).

From this we can deduce two facts: (i) Argumentation strategies change depending on the context. Confronted with the public, a representative of a research body must be able to show some results — not just considerations fraught with uncertainties (see Krohn 1986:24—26). (ii) the choice of the selected argument is not made in an accidental manner. It depends on the environment, and on other groups of actors who might be helpful in pushing the arguer's own interests (which actually might be served in different ways). Thus the hints given by Jordan concerning a future military space policy (as the DFVLR has a good deal of experience with military research) possess a certain logic.

Political actors are using the scientific arguments given by the DFVLR as sources of legitimation; in return, they furnish this institute with political concepts such as "autonomy," "leadership," or "space defense," which serve as legitimation for certain types of R&D. It is thus that the DFVLR's lead in the space sciences can be attributed to the fact that it is part of a network.

Politics

In politics we can see a similar development. In the 1970s a major decision was taken in German space politics: to concentrate on the commercial use of space flight and to take the first steps toward manned space flight (see Büdeler 1978:25, 61). But when discussion about the Strategic Defense Initiative (SDI) and the space station emerged, Research Minister Riesenhuber was not really willing to join those projects because he couldn't see any R&D-like justification (13): he preferred to continue unmanned space research or to construct a much smaller station than Columbus (14). Riesenhuber had no allies in the field to push through that point of view except some "outsider" scientists (15) or the Green and the peace movements. The contrary view, however, gained strength during the SDI-discussion (see Weyer 1986) that helped to minimize the resistance to space politics (even military) and brought about a specific European interpretation of space technologies as keys to the future (16). Baden-Württemberg's Minister President, Lothar Späth, is one of the spokesmen of that viewpoint, and he characterizes the space and defense technology industries as the "pulse generator for technological progress as a whole" that a modern, competitive economy must not miss (17). His argumentation actually continues a trend in former space-research policy, which since the 1970s has been characterized by an economic orientation — that is, a direct support of the space industry (see "Faktenbericht," pp. 91—93). But now the argumentation has shifted, for (i) defense issues are regarded as equivalent, and (ii) the fall-out benefits from space flight, formerly anticipated to accrue to a small sector of the space industry, are now extended to the industry as a whole.

This reference of space politics to international economic competition, to defense problems, to international obligations (concerning ESA and the transatlantic connection), and to the "future" have made it difficult for Minister Riesenhuber to resist the trend (which may in a few years become a serious threat to his research policy [18]). The field of "research policy" thus has been restructured by the recent space decisions, which did not follow the internal logic of the field but were based on arguments imported from outside (international relations, defense policy, economic). And the relations within the larger field of politics have been rearranged too — for example, in favor of defense and to the disadvantage of détente policy.

In public statements, Riesenhuber finally took over the arguments in an apodictic manner that formerly was uncharacteristic of him (19). The restructuring of research politics has not yet been completed, however. The recent discussions about the foundation of a German space agency, or even a space ministry (20), indicate that there is still a strong drift toward a closer connection of politics with space issues. This would probably weaken other areas (such as the BMFT) to the advantage of the hybrid space community.

The Space Industry

The space industry — which in Germany is almost identical to the technologically advanced parts of the defense industry, with significant overlaps with the nuclear industry — naturally has primarily economic interests, and thus some interest in maintaining a suitable political milieu in the future as well. But such a forecast doesn't extend the planning-horizon to more than about five years (see, e.g., Steinecke 1986); every measure that gives returns only after ten, fifteen, or more years is viewed as irrelevant to industry's interests. So it can be understood, that the cost of the space program must be paid by governmental budgets. On the other hand, the space industry has built up remarkable capacities in terms of highly trained staff and technically sophisticated apparatus, which (after having finished former contracts) call for follow-up actions on an ever-higher level, especially in those industries that operate nearly exclusively under governmental contracts. Since almost no relevant private market exists for the products of the defense, nuclear, and space industries, they are inevitably dependent on steady governmental support, which could better be allocated by long-term programs than by sporadic measures (see Steinecke 1986).

But the space industry is only a very small (although powerful) sector of German industry, with only about 4,500 people employed (Weltraumforschung, p. 35). Even at Messerschmitt-Bölkow-Blohm, which as a whole is dependent on governmental support at a rate of 80% (Büdeler 1982:66), the space sector covers only 4% of the group (p. 71). Therefore the steady promotion of the potential benefits of space technology for the total economy and for the future of the whole society is a crucial tactic in maintaining the space industry's public image (see BDLI 1984:6, 14). And the phantom of a so-called user-industry waiting for the opportunity to utilize space for production, and so on, also plays a great part. The argumentation strategies of the aerospace industry are fixed around one essential point: avoiding the application of the internal logic of economics to the field (that is, the market laws). A memorandum of Bundesverband der Deutschen Luftfahrt-, Raumfahrt- und Ausrüstungsindustrie (BDLI) (Federation of German Aeronautics, Astronautics, and Equipment Industry), released in 1984, argues in that way and calls for permanent subsidies (BDLI 1984:46), for protectionism (p. 44), and for a nationalization of the market (p. 31). It is openly admitted that the commercial market is not sufficient. Consequently, the BDLI calls for political measures that could open the European and the world markets as well as the home market. Frequent remarks about the necessity of a military use of space are part of this strategy.

Here again, we can see externally borrowed arguments working as amplifiers for one's own interests, which have been transformed in a specific way. It is rather striking that K. H. Allgaier, top manager of MBB in the field of defense initiatives, declares: "For me the threat becomes more critical than could be expected by the Bundeswehr plan for the next fifteen years" (21). And one of his recommendations is to build up a European ATM (Anti-Tactical Missile Program) combined with an early warning system consisting of about 100 reconnaissance satellites (22). It is obvious that it is not "military needs" (WEU 1984, 1:14) but primarily industrial interests that are pushing the formation of a military space market. And there is hardly any other actor on the stage (except the WEU) who draws attention to the military use of existing of future space technologies in the openhearted manner of the space industry (see BDLI 1984:40, 45, 49). In public, however, these arguments are hidden behind other scenarios, such as environmental protection, development aid, weather forecasting, securing the future, and so on (23).

This transformation of the immediate interests of the space industry into the language of economics and defense again reveals the mechanism, that arguments borrowed from external areas help to redefine the problem of the space industry (making a profit by being the technologically most advanced player in the competition game) in terms of the problem-perception of another actor — a mechanism that imports credibility from outside and together puts pressure on this actor, in that he loses a part of his legitimacy if he cannot show a solution to the problems defined in this way. And the mechanism also explains the political impact that a very small sector of the high-tech industry has gained. Finally, it can be proved that the German aerospace industry is a major driving force in the process of bringing about manned space flight on the one hand, and military uses of space on the other.

The Military

One remarkable fact that emerges when we compare the German and the U.S. or French situations, is the total absence of the military in the public debate about space flight. This is mainly due to the special situation of West Germany after 1945, which hampered the continuation of a militaristic policy. In the course of time, the peaceful image of West Germany made itself independent in some way and began to form a distinct part of traditions that — for example, in research policy — are difficult to bypass (see Haunschild 1986:61). But the intense debate launched among historians in 1986 shows the attempts of some groups to change the conception of history and the attitude toward imperialistic and militaristic politics — a process that is intended to make Germany equal (e.g., to the French) in terms of freedom of action in every field of politics.

While direct intervention of the military into the field of space flight is very restricted, it has to be considered that:

(i) There is a long tradition of military involvement in space flight (24).
(ii) The German military has a far-reaching influence on the civilian space program.
(iii) Plans of nonofficial organizations or individuals, but also of the well-authorized Western European Union, for a future military use of space in no way have to stand outside the international state of the art.
(iv) Many suspicions may be raised about the civilian use of the space programs of the European Space Agency (ESA) and the German Research Ministry (BMFT) (25).

The present situation in German space projects is characterized by different ways of informal incorporation of the military into the decision-making process on the one hand (IABG 1986:8), and by an evident orientation of space R&D toward the so-called "utilizing body" BMVg (Ministry of Defense) on the other (DFVLR 1986:5/2). The latter can be found even in cases where the military doesn't pay for the research projects (see DFVLR 1986:1/1—6, 9/2). And it is a remarkable tendency to be discovered in the DFVLR budget that the civilian BMFT totally finances the (partly international) basic research — for example, in microwave remote sensing (DFVLR 1986:6/5) — while the military application of that technology is exclusively BMVg-financed (*ibid.*:6/2, 9/2). Such indications may raise the suspicion that civilian programs, especially when conducted in international, peaceful cooperation, are at least closely coordinated with military interests. The IABG (Industrieanlagen-Betriebsgesellschaft) — one of the government-related brain trusts — provides us with a logical explanation of these facts: "A further reason for the restrictive attitude of the BMVg may be seen in the fact, that the announcement of a demand within the BMVg would lead to the announcer's taking over the costs to the total extent" (IABG 1986:30).

As space projects do not possess the highest priority within present "Bundeswehr" planning, and concrete military requirements transformable into research or procurement orders are not at hand (see IABG 1986:8, 30), it seems to be reasonable that the military remains in a waiting position as long as the technologies developed by BMFT and ESA do not drift in a direction that would be useless to the military. There is a prominent reason for the absence of the military in the discussion: The frame of reference of the space debate — be it the U.S. or the French space program — is fundamentally military-oriented, so that the military as an actor need only come into play if the plans of the other groups involved deviate substantially from copying, for example, the U.S. model. In this way the German military profits from the military character of the space discussion elsewhere. Why should the military risk direct intervention into space projects if even the planned space station is connected in the DFVLR plans with the "utilizing body," the BMVg? (26).

On the other hand, there exists a specific German interest in a so-called extended air-defense (also named European Defense Initiative, EDI), which is not only SDI-related, but also necessarily partly space-based, even if the defense minister hesitates to admit that in public. His undersecretary, Lothar Rühl, knows the correlations very well (see BMVg 1986:59—60, 63); he is also one of the authors of a memorandum that outlines a future military involvement in space from the perspective of Germany as a new superpower (see section III, below). But apart from these future scenarios, which sometimes do not conceal their references to certain political traditions or their radical militarism (27), the international military integration plays a decisive role in future space-defense politics. It is not only, as mentioned above, the SDI program that made access of the military in Europe to outer space more acceptable: the BMVg also participates in the NATO satellite systems Satcom and Navstar (IABG 1986:8, 30; see also Althainz et al. 1984, and Scheffran 1984 and 1985).

The most prominent step toward the creation of an independent European space power was the revitalization of the WEU and the dealing of that body with military space matters. In 1984 a recommendation of the WEU was passed calling for "a defensive European military space programme" inside "an institutional framework untrammelled by the political inhibitions of the ESA convention" (WEU 1984, 1:2), but nevertheless making use of nearly all the ESA-founded space technologies, such as ERS-Statellite (microwave remote sensing), Meteosat (weather satellite), space station, shuttle Hermes, and so on (pp. 3, 14—15). The report given to prepare the WEU decision passionately argued against the hope of a "considerable technological spin-off" (p. 14) from the space station and called instead for a military use of space, which would make the enormous amount of money to be spent more reasonable. The report also showed that priorities should be given to the following four elements: (i) telecommunication satellites (soon available), (ii) a military observation satellite system (prototypes in construction), (iii) a navigation satellite system (planned), and (iv) attack satellites (no priority) (pp. 13—15). Antisatellite weapons, large-payload launchers (Ariane), and the shuttle Hermes were added to the list of wishes of the military, too. Here again it can be seen that ESA technologies are at least compatible with military interests, and the military is now going to lay claim to those — all civilian-financed — projects.

It can only be speculated that the German military plays a part in these plans. But the minimum that can be concluded is this: The frame of reference for the decisions of the military has seriously changed since Star Wars, and space defense is on the agenda. It has become increasingly easier to demand certain types of space techniques if such demand can be supported by establishing links to other reference groups.

Hence we can analyze this actor, too, in the same way as the others above. The changing ambience has deeply influenced the structure of the field to the extent that space defense is now a matter of political debate. The support from outside (especially by means of international military cooperation, as in NATO and the WEU) has strengthened — or is going to strengthen — those elements within the military who put emphasis on future space-based defense scenarios. The mutual relations with other groups of actors are consolidating (which means the network grows and becomes tighter), so that those actors now can refer to a military demand that helps to legitimate political strategies and — above all — industrial requests.

Conclusion

In reviewing the German space program and the actors engaged in it we have a situation similar to that which other analysts of technological dynamics have previously found (see Kitschelt 1979, Radkau 1983 and 1986): There is no clearly identifiable steering-center; instead, there is a sum of actors with initially different interests, merging together by processes of mutual reinforcement of their positions and interdependencies in a way that can be shown as a translation of arguments into the language and terms of the partner, thus initiating new problem-perceptions on the one hand and borrowing legitimacy on the other. The mechanism analyzed works in two directions: first, by rearranging the frame of reference the co-players make use of when reaching their decisions; and secondly, by using the external environment as a resource that supplies the power to restructure the field one belongs to (these being two ways of viewing the same mechanism). In the final section of this paper the idea outlined here will be discussed again.

But first it is essential to look at the hybrid structures and the new institutional settings emerging in the space field. We will take a memorandum released in 1986 by the Deutsche Gesellschaft für Auswärtige Politik (DGAP) (German Society for Foreign Policy) as an indicator of the state of the discussion inside that hybrid structure. The DGAP memorandum unites representatives from all the areas mentioned above; it can be viewed as one attempt to harmonize the different positions and to find some sort of common phraseology, which is indispensable for a confrontation with the public outside the space lobby. (For localization of the DGAP in the field, see figure 2.)

III. German Space Plans and Their Financial and Social Costs

The technological and political aims of the West German space lobby are frequently described by representative bodies of this group that exert an intermediary function and thus form a sort of inner core. They can be located in political research institutes, foundations, and other organizations of that kind, which are not (or not primarily) part of one faction. These hybrid organizations usually not only serve in coordinating the space lobby, they also aim at influencing public opinion. They thus serve an important function for the public relations of the space lobby that exceeds that of the individual inner-systemic actors. It is important to mention that the legitimation of the space lobby's requests with reference to the environment (society, public, politics) requires other forms of argumentation strategies than those needed in the inside relations between the different groups — a phenomenon already mentioned above.

One central demand to be found in this context is the goal of making Western Europe a "space power of the twenty-first century" — a demand that calls for an enforced European and German "engagement in space" (DGAP 1986:43). Though Germany is working toward these aims within a European structure, it becomes evident that such cooperation has only a tactical importance for the final goal: "to realize West German space power *by means of* a European cooperation" (*ibid.*, emphasis added). Within the frame of the European Space Agency with its already advanced rocket program Ariane, the planned European contribution to the U.S. space station Columbus, and the 1986 predecided plans for building a minishuttle called Hermes, Germany shall — following the words of the space lobby — play a greater part in deciding, financing, and performing future projects.

The self-perception of the lobby — as among others presented during a parliamentary hearing in 1985, or in the DGAP memorandum "German Space Policy at the threshold of the Century" (released in 1986) — about the importance of the period of the mid- and late eighties is that West Germany needs a long-range political program based on a "reorientation of German space policy" (DGAP 1986:38), if the country does not want to risk losing its political and economic position in the world and if it wants to profit from the military potential of space flight. In the words of the cited memorandum, space activities enable the countries engaged "to strengthen substantially their political and economic influence and thereby their position within the international community"; and some sentences later, with greater clarity: "At the bottom of that race for predominance in space are . . . primary geopolitical and military profits" (DGAP 1986:21).

This basic assumption, that Germany is forced to participate in the space race in a more intensive way and in a more competitive manner, leads to the following consequences:

(i) Europe has to gain "autonomy" (Anhörung, p. 112) in space. This means getting rid of the dependence on U.S. space flight, which has sometimes been shamelessly exploited by the United States in the past. As the principles of space flight are the same for all countries, Europe must install a perfect copy of the U.S. technical configuration; this means constructing the "triad" of Ariance (the launching system), Hermes (the transport system for manned missions, too), and Columbus (the space platform in a lower orbit).

(ii) Europe must change direction and enter the sphere of manned space flight. Though everyone knows that "space sciences would prefer unmanned stations or space probes," manned systems are estimated to be irreplaceable because of their "higher political symbol — effect" (DGAP 1986:38). Even if there exist no more arguments for manned space flight than national identity and international symbolic competition, the entire European space concept and its enormous costs are centered on "man in space."

(iii) If geopolitical power will depend in future not only on the possession of atomic weapons, but more and more on the possession of space weapons, then participation in the space race is unavoidable for a country that is defined — at least by the space lobby — as a future world power. The memorandum thus calls for an "adaptation of the defense concept to technological development in the East and West" (DGAP 1986:42), as a part of the notion that West Germany should be "a shaping force of the Western European space power" (*ibid.*). The demand "partially to correct the political power imbalance between the superpowers and the European states . . . in the area of space policy" (p. 22) shows a policy-conception, underlying the space program, whose fixed points are the correction of the postwar political constellation and the reentry of Germany into the club of superpowers (28).

The realization of this concept means — and the space lobby is well aware of this — a far-reaching reorientation of West German and European space policy. Both have in former times focused on peaceful space research, with an emphasis on such scientifically useable and commercializable projects as, for example, the most successful rocket Ariane. But the described aims require a fundamental shift in program structure and research priorities, in technical configuration, and — last but not least — in the financial volume of space research. The recent decision for the new Ariane V shows that the ESA program is headed in the direction promoted by the space lobby (29).

The decision for the space trio of Ariane, Columbus, and Hermes can in no way be compared with any previous decision for other big technologies (30). The amount of money required is hardly foreseeable, and the financial consequences will probably lead to sociopolitical measures and to reorientations of research priorities that are even now (before any social opposition has formed) being anticipated and intensively discussed by the space lobby. Let us look at the costs:

It is assumed that the German share of the R&D costs of all three projects (31) will be about 6.5 billion DM, to be paid within about eight or ten years. It results from the following items: The development of

Ariane V will require, up to 1994/95, about 7.5 billion DM, with a German share of 22% (1.65 billion DM); for Columbus the same amount has to be spent, but Germany will contribute 38% (2.85 billion DM). The fulfillment of these obligations already requires a raise in the German space budget from 1.1 billion DM (1986) to 1.6 billion DM (1992), which is frequently demanded by representatives of the space lobby (32). There are even some who insist on a "doubling of the space budget within the next ten years," thus reaching the U.S. rate of 55 DM per inhabitant per year, or in other words, a German space budget of about 3.3 billion DM (33).

But the story doesn't end here; there is still Hermes to be paid for (34). It is estimated that Hermes will cost about 5 to 6 billion DM , but insiders know that this assessment is much too low. As the well-informed *Frankfurter Allgemeine Zeitung* states, the experts "even today calculate with sums twice as high *or more*" (25 August 1986; emphasis added). If the costs were to remain constant (which nobody expects), Germany would have to pay, at the rate of 30 to 33%, the sum of about 2 billion DM for Hermes — which means a total contribution of about 6.5 billion DM to the Western European space program. No one knows where to acquire such funds, and when the members of Parliament asked the lobbyists during the hearing in the German Bundestag in 1985 they got neither answers concerning the total amount of money to be spent, nor any sign of willingness on the side of industry to contribute financial support. The lobbyists simply refused to give answers or side-stepped (Anhörung, p. 205), as shown in the following example: "It requires time and costs money. The efforts are worthwhile, however" (Anhörung, p. 233).

A very popular argument — well known from the U.S. shuttle — is the promise of reducing space transport costs in some twenty or thirty years (see Anhörung, p. 246) which is one of the major arguments to support Saenger (35). But the fact is that the total system will cost West Germany *at least* 6.5 billion DM, or one entire annual research budget (which in 1987 is at a level of about 7.5 billion DM). This fact makes space technology an absolute novelty in the history of German research policy; even the fast breeder, when contracted in 1972 (with total costs estimated at 1.54 billion DM), demanded "only" half of the annual research budget (1973: 3.14 billion DM) (see Keck 1984:206).

We also have to take into consideration the fact that Hermes — and probably the other components of the system, too — will effectively cost twice as much or more. If we characterize space technology as being in many ways an unpredictable line of technological development similar to the fast breeder (36), and if we take into consideration the cost explosion of the German breeder from 310 million DM (first plans in 1961), to 1.535 billion DM (contract in 1972), to 6.05 billion DM in 1982 (see Keck 1984:203—208) and some billions more in 1986, with a gradient factor of about 5 (from 1972 to 1986), we can easily produce three alternative scenarios of estimation and assessment of the sums to be spent for the space program. This will enable us to draw some conclusions concerning further research policy.

Tables 1 and 2 show the development of the German space research

TABLE 1
Development of the German Space Research Budget
(compared with the total and the nuclear research budget)

Year	BMFT (total)	a.i.	Space R&D	a.i.	Nuclear R&D	a.i.	Space % BMFT	Nuclear % BMFT
1973	3,140		600		770		19.1	24.5
1974	3,680	17.2	590	−1.7	810	5.2	16.0	22.0
1975	4,080	10.9	580	−1.7	1,200	48.1	14.2	29.4
1976	3,960	−2.9	650	12.1	1,310	9.2	16.4	33.1
1977	4,210	6.3	610	−6.2	1,410	7.6	14.5	33.5
1978	4,950	17.6	660	8.2	1,530	8.5	13.3	30.9
1979	5,570	12.5	730	10.6	1,650	7.8	13.1	29.6
1980	5,840	4.8	790	8.2	1,810	9.7	13.5	31.0
1981	6,070	3.9	800	1.3	1,980	9.4	13.2	32.6
1982	7,080	16.6	880	10.0	2,700	36.4	12.4	38.1
1983	6,920	−2.3	900	2.3	2,570	−4.8	13.0	37.1
1984	7,050	1.9	900	0.0	2,620	1.9	12.8	37.2
1985	7,200	2.1	960	6.7	2,650	1.1	13.3	36.8
1986	7,410	2.9	1,111	15.7	2,470	−6.8	15.0	33.3
1987 D	7,560	2.0	1,261	13.5	2,323	−6.0	16.7	30.7
Inc. 1983—87 (average)		2.2		9.0		−2.4		

a.i. = annual increase; all references in million DM.
(Source: Bundeshaushaltsgesetze 1973 passim)

TABLE 2
Linear extrapolation of the German Space Research Budget up to 1995 based on the average increase 1983—1987
(compared with the total and the nuclear research budget)

Year	BMFT (total)	a.i.	Space R&D	a.i.	Nuclear R&D	a.i.	Space % BMFT	Nuclear % BMFT
1981	6,070		800		1,980		13.2	32.6
1982	7,080	16.6	880	10.0	2,700	36.4	12.4	38.1
1983	6,920	−2.3	900	2.3	2,570	−4.8	13.0	37.1
1984	7,050	1.9	900	0.0	2,620	1.9	12.8	37.2
1985	7,200	2.1	960	6.7	2,650	1.1	13.3	36.8
1986	7,410	2.9	1,111	15.7	2,470	−6.8	15.0	33.3
1987	7,560	2.0	1,261	13.5	2,323	−6.0	16.7	30.7
1988	7,726	2.2	1,374	9.0	2,267	−2.4	17.8	29.3
1989	7,896	2.2	1,498	9.0	2,213	−2.4	19.0	28.0
1990	8,070	2.2	1,633	9.0	2,160	−2.4	20.2	26.8
1991	8,248	2.2	1,780	9.0	2,108	−2.4	21.6	25.6
1992	8,429	2.2	1,940	9.0	2,057	−2.4	23.0	24.4
1993	8,614	2.2	2,115	9.0	2,008	−2.4	24.5	23.3
1994	8,804	2.2	2,305	9.0	1,960	−2.4	26.2	22.3
1995	8,998	2.2	2,513	9.0	1,913	−2.4	27.9	21.3
Inc. 1983—87 (average)		2.2		9.0		−2.4		

a.i. = annual increase; all references in million DM.
(Source: Bundeshaushaltsgesetze 1981 passim)

budget in the past and a linear extrapolation up to 1985, based on the average increase rate of the last five years. Table 3 and figure 3 try to assess the future development of the space budget: scenario A supposes constant prices, scenario B assumes a doubling of costs, and scenario C hypothesizes an explosion of costs by a factor of 5. All three scenarios entail attempts to manage the growing problems by redistribution of funds within the research budget. There is also the underlying assumption that funds have to be spent, not constantly, but with a maximum in the years 1992/93.

With the help of these three scenarios the problems of the German space commitment can clearly be shown. Even the sums resulting from scenario A could be managed only by a constantly increasing space

TABLE 3
The West German Space Research Budget up to 1994 (Estimation)
(Three alternative scenarios)*

Year	BMFT	Scenario A			Scenario B			Scenario C		
		NSP	SP. R&D	%	NSP	SP. R&D	%	NSP	SP. R&D	%
1986	7,410		1,110	15.0		1,110	15.0		1,110	15.0
1987	7,560		1,260	16.7		1,260	16.7		1,260	16.7
1988	7,726	300	1,500	19.4	500	1,500	19.4	500	1,500	19.4
1989	7,896	700	1,700	21.5	900	1,800	22.8	1,500	2,400	30.4
1990	8,070	1,000	1,950	24.2	1,800	2,650	32.8	3,000	3,700	45.8
1991	8,248	1,000	2,000	24.2	2,300	3,000	36.4	5,000	5,500	66.7
1992	8,429	1,300	2,200	26.1	2,600	3,250	38.6	7,500	8,000	94.9
1993	8,614	1,300	2,300	26.7	3,200	3,700	43.0	9,500	10,000	116.1
1994	8,804	1,000	2,100	23.9	1,700	2,300	26.1	5,500	6,500	73.8
sum		6,600			13,000			32,500		

BMFT = Bundesministerium für Forschung und Technologie; NSP = New Space Projects (Ariane V, Columbus, Hermes); SP. R&D = total Space Research Budget; % = Space R&D share of the total research budget; All references in million DM.

* For graphic illustration, see figure 3.

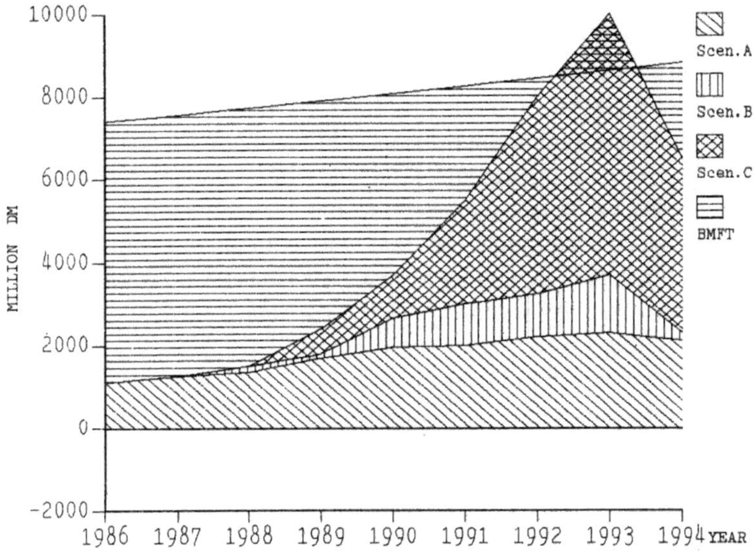

Fig. 3. Assessment of costs of space-scenarios.

budget; scenario B can only be performed by a new distribution of funds both within the space budget (to the disadvantage of other programs, such as space science), and within the entire R&D budget (to the disadvantage of such areas as health, ecological, social, and other research). But scenario C — which may be the realistic one — shows how the space budget will "eat up" the R&D budget from the inside, thus provoking a search for other sources of financial support. Herein lies the major force affecting future German research policy, even if the entry into space flight is less spectacular than in the United States (or in science fiction). Given the case of an explosion — either of costs or of the launching system — the hold of other projects with more reserve capital on the space program is practically preprogrammed, which would probably lead to an altogether new strategy of space flight and space research (37). And as has been shown above, these considerations are no mere speculation.

We can conclude, simply by looking at the financial situation of the space program, that a shift in balance from civilian to other types of space research (which may be military) and a massive burden on the federal budget are preprogrammed. If they come about, these incidents would create an important restructuring of the whole field. The space lobby is well aware of these problems, as the DGAP memorandum points out: "An essential intensification of Germany's space engagement will make sensitive displacements in the federal budget necessary . . ." (DGAP 1986:39). Because the decisions of the mid-eighties are regarded as fundamental determinations that will commit a great amount of money and research personnel to space activities for a period of at least fifteen years (see Anhörung, p. 204), the space advocates see quite clearly that the government "would be forced to cancel considerable sums in other relevant political areas" (DGAP 1986:43).

To avoid foreseeable political trouble, the DGAP memorandum recommends that the German population should be convinced of the necessity of space flight (DGAP 1986:39—40) and that the space lobby should concentrate its political activities on the public "legitimation" (*ibid.*, p. 13) of space policy. Here we can see an interesting example of plans to use public debate as a source of support for the space program, and of well-thought-out ideas of modern sociomanagement aimed at convincing people to favor a policy they would normally resist. The space lobby is well aware that the management of their outside relations is a crucial point for succeeding with their politics, and that organizations specialized in public relations are required. But the major aim of the lobby is to keep the ongoing discussions on a level that does not advance to a principled jeopardizing of the space program.

Because it is obvious that, once they are fixed, decisions have a kind of irreversibility and self-dynamics, the principal appointments should — according to the space lobby — be made now, programmatically, without any technology assessment. Or, in the words of the research minister: "It will later be the task to decide whether the expenses were profitable" (38). The motto "decide now, and think about it when nothing can be discussed any more" is very popular in the debate. But at the same time one is conscious of the fact that such a mechanism cannot work forever. Similar to U.S. models of multiyear authorization, the German space lobby is attempting to decouple the space program from public control on the one hand, and from incalculable discontinuities in the democratic and parliamentary procedure on the other. The president of the Berlin Technical University, Manfred Fricke, declared in his opinion before the parliamentary committee: "Nothing is more harmful to a long-term protection of existence than turning only towards partial solutions oriented to election dates" (Stellungnahmen, p. 7). Space scientist Klaus Pinkau of the Max Planck Institute reflected publicly about the incompetence of the research ministry to plan and to carry out the space program (Anhörung, pp. 124—125) — reflections that led to the idea of a national space center by strengthening the DFVLR on the one hand, and by a partial reorganization of the European Space Agency (ESA) on the other hand, in such a manner that public controls on both levels are diminished (DGAP 1986:54—55; Anhörung, p. 158). Here, as well as in the budget discussions, we can see the space lobby busily trying to extend the area of their influence into other fields.

Loud voices are proposing that space policy should be nationalized (against ESA) while at the same time the space program should be carried out under a new authority that is not obliged to the research policy of the Research Ministry alone, but is also open to other interests, which — following the proposals of space advocates — should be military, too (39). As Dornier manager Helmut Ulke points out, "more ministries should engage in space flight (postal services, defense, interior, development aid)"; he continues: "It would be necessary to build up a German space agency to conduct these tasks of coordination and goal formulation" (Stellungnahmen, p. 9). In the same way, the DGAP memorandum calls for other ministries to engage in, and to pay for, German space activities (DGAP 1986:56) — though there can be no doubt that it is the defense ministry that is primarily meant.

There exists, thus, a large amount of evidence for potential future military involvement in space flight, which indicates that the center of gravity within the field will be shifted in the direction of the military (see figure 2). This analysis also shows, however, that the public is a crucial factor that shapes the strategy and tactics of the space lobby in an important manner. At times one truly has the feeling that the lobby is playing a game with the public, as when one reads: "... confronted with the public ... one should point out in the first instance the economic and social advantages" of space flight (DGAP 1986:40) – a statement to be found only a few sentences after explanations of the redirection of funds in the federal budget to the disadvantage of social, health, ecological, and other issues.

The next section will discuss some recent legitimation strategies and will include considerations about the relation of civilian and military use of space, because problems of legitimation can arise only if there is a barrier between two areas (consisting of the lack of direct connections or applications) that must be bridged argumentatively.

IV. The Problem of Public Legitimation and the Military Use of Space Flight

Two facts must be borne in mind. First, German research policy is aiming for development of the same technologies as the SDI program, because these technologies are estimated to be the key for the future; this assertion applies equally to the European EUREKA program (40). The frequent hints that Europe has to counterbalance the U.S. efforts (which are carried out within a military program) by its own research activities (41) point out this close connection of research goals. The only clear difference lies in the fact that any sort of concrete goal-definition comparable to the "construction of a strategic missile defense" is missing in German and European programs, which thus seem to demand the development of rather unspecific technologies.

Secondly, there no longer exists a general resistance to the publicly pronounced promotion of military research. As R&D Undersecretary Hans-Hilger Haunschild has pointed out, he sees "a clear separation" of civilian and military research, but his reference to the equality and interchangeability of both "fields of responsibility," as he calls it (1986:61), shows a remarkable upgrading of military research, which can thus be considered as legitimate as civilian research. Former barriers are being dropped, as Research Minister Riesenhuber did when he declared that EUREKA-sponsored research "could be applied in the military sector, too" (42). We find this trend also in space research, where the lobbyists complain that the situation of German industry in international competition is "today unbalanced, since only West Germany . . . doesn't carry out military space projects" (43) — and yet, at the same time, there exist detailed plans for a European Anti-Tactical Missile Program (ATM) (DGAP 1986:30, 48—49), which will lead to a European Defense Initiative (EDI) that can be combined with SDI (44).

Besides a number of hidden hints, we can find a frank and open-hearted description of the military character of the European space program given by Ernst Högenauer, a leading scientist of the biggest aerospace trust in Germany (Messerschmitt-Bölkow-Blohm), in a widely circulated public magazine — a procedure that has some of the character of launching a test-balloon to explore the public acceptability of military space plans. Högenauer points out that hypersonic space-crafts like the European shuttle Hermes are of great importance for both military and civilian aeronautics, thus focusing not on the direct military use of space technology but on a celestial spin-off for terrestrial purposes. Concerning military fields of application he writes: "For military aeronautics, hypersonic planes represent the perspective for the next generation of combat fighters"; these fighters would be able to reach "any point of the world within one hour," to "move above a low orbit" or to "stay for longer periods combat-ready in this orbit" (45). On the other hand, the perspectives for a civilian hypersonic transport plane are so vague that even Högenauer admits:

> Though the civil application of hypersonic technology may seem possible only after this development has been sufficiently qualified by space flight and the military, the civilian concept has at least got an adequate name for the present: Orient Express (46).

There exist, as the Concorde disaster shows, no reasonable civilian applications of that technology, and probably there never will be any (despite the weak comfort of giving the "project" a name) — and the cited representative of the military-industrial complex is fully aware of this fact. The problem is: How can we justify plans of this kind?

It is my hypothesis that space flight nowadays is no more than the continuation of a former exclusively military technology with a new civilian label, a sort of (additionally "invented") external defense belt (47) constructed around the hard core of military space flight — thus protecting military purposes from public inquiry and broadening the ways of acquiring funds for this very expensive research and technology. This switch in labeling and legitimation (but not in fields of application, as I will show below) was accentuated by Undersecretary Haunschild, who characterized hypersonic research as a "classic field of activity of military research" (1986:62). It was only after the discussion about a European shuttle, Haunschild believes, that another field of application emerged (he avoids the term "civilian"!), though it is still "unproven" that anything in this sphere makes sense in economic terms. But the following sentence is remarkable: "Obviously the primary foundation for such technologies can reverse to defense technology" (*ibid.*).

Here the dual-use thesis in its latest version — the assertion of the interchangeability of application and legitimation — is used to establish the equality of both types of research as different means to serve the common goal: the development of high technology. If a former military technology can be pursued in a civilian context, then — according to this thesis — it doesn't matter in which area the research is done. This thesis has replaced the untenable spin-off thesis during recent years, but it is based on the premise that underneath the interchangeability of legitimation strategies (a case whose functioning still has to be explained) there exists a factual interchangeability of technologies between the different fields of application. This thesis only works in our context because space flight is regarded as a civilian purpose, but the suspicion arises that military technology has found a convenient label — especially in a country where military research was forbidden for a time (48).

If, for the time being, we accept the assumption that space flight is a civilian undertaking, then it is our task to ask for reasonable projects in space flight and space research that will support civilian interests; these interests might be primarily economic and scientific, but could be cultural, ecological, social, and international-relations as well. If you ask the space advocates to deliver civilian-based arguments for space flight, you will very quickly be confronted with the fact that there are none, except for some vague hopes and speculations that are in no way comparable to the very concrete plans for military use – for example, for a European ATM Defense System (EDI).

This is the lesson that the members of the German Bundestag learned when examining the arguments of the space lobby in 1985. Without reproducing and commenting on the whole affair here at length, the core argument is: There is no convincing argument for entry into manned space flight, but just *because* there is "no logical evidence" (Anhörung, p. 142) we should take exactly this risky step to get rid of our knowledge deficit. The space scientist Erhard Keppler called this argument (which was nonetheless also his own) a "logical crack" (*ibid.*), to follow which one required a whole portion of irrationality and metaphysics.

The best way to answer the question about potential civilian uses of space technology is to ask the carefully calculating industry whether an aerospace firm would enter the field of manned space flight and make investments in the hope of getting a profitable return. The answer again is disappointing: neither the U.S. nor the German aerospace industry spends a penny on any space project in which the government does not invest more than one-and-a-half times as much, and the nonspace industry has no interest at all (49). In the German D-1 mission in 1985, only three experiments came from German industry. During the parliamentary hearing, the MBB representative Othmar Heise presented the calculations of industry as follows: "Space flight . . . is a technology that upon application today cannot be said to yield a sufficient return on investment" (Anhörung, p. 206).

At this point one cannot identify any civilian application of manned space flight. For scientific purposes man is only an interference factor, as James Van Allen pointed out very convincingly (50). The economically profitable sector of space nowadays is the geo-constant orbit, which can be reached by "simply," "cheap," and unmanned throw-away rockets available today (51). Material experiments in space laboratories still have few or disappointing results (Anhörung, p. 178), and even if success is attained there will be no prospect for removing the motor industry to space, as the critical scientist Hans-Peter Dürr remarked (Anhörung, p. 180). Not least important: celestial spin-offs for civilian applications, such as the successful Dornier Nierenlithotripter, are so scarce that they cannot justify the enormous costs of today's space flight, nor those of future projects. Further, these spin-offs mostly come into existence by pure chance, like the new material for an artificial hip-joint that came out of Tornado aircraft research (see Abstein 1986).

One potential objection against these arguments is that in some fifteen or twenty years manned space flight might be commercially viable, too, thus following the example of the Ariane success (see Junker 1986) — but this objection is valid only if you exclude from the calculation the billions of dollars spent before (which governments usually do, as with space flight or nuclear energy). Under the conditions of an independent cost-benefit analysis, enterprises (German as well as U.S.) are simply disinterested in the utilization of space (52). There is no civilian, commercial, or scientific demand for manned space flight, even if in some unforeseeable future an artificially produced "market" (which is really a balance of different modes of subsidies) may emerge.

Another potential objection might be the question of military uses of manned space flight. Obviously all military missions may be performed by automatic systems, and such ideas as the use of the shuttle for a battle-observation and battle-management platform seem to be a bit antiquated in times of worldwide data communication. But there are some reasonable arguments for the military use of manned space flight (which are at least more reasonable than the arguments for civilian use):

(i) During the construction, installation, and testing of new automatic weapons systems in space, man could be a help.

(ii) It is unavoidable for the military to keep up with the state of the art in the field of weapons technology, which means being prepared for the future option of a new battlefield in space.

(iii) Research done in space projects has several applications (spin-offs) in the field of so-called conventional military technology — that is, in constructing new "intelligent" weapons with novel features and new missions that might be superior to the enemy's capabilities.

(iv) Last, and not least: even the military profits from the symbolic effect of manned space flight and from the public enthusiasm and willingness to pay for rather expensive space projects.

We can summarize here that it will remain a difficult task for the space lobby to justify all the planned space research in view of the lack of civilian uses. The public seems to be a crucial frame of reference for the strategies of the space lobby that determines their actions in a certain way — for example, forcing the lobby to increase the acceptability of space projects by relabeling them. On the other hand, we see the lobby's attempt to influence public opinion and decrease the general resistance to military plans (which had been increased by the so-called Nachrüstung with Pershing 2 and cruise missiles). Here we can see an example of the "struggle for borders" (Krohn and Küppers 1987:5) taking place between different systems and their respective environments.

V. Conclusion

In this paper an attempt has been made to analyze the emergence of a novel space technology (manned space flight) and to interpret current changes in R&D policy in West Germany by using elements of the concept of self-organization. The so-called space lobby has been described as a social structure that emerged through the interplay of (for the present) quite different interest groups; parallel to the generation of a new production (a novel big technology), this structure gained coherence and obtained a specific stability and self-dynamics, thus producing irreversibilities. The mechanism that brings this hybrid community together can be analyzed as a complex process of adjustment of interests and transformation of arguments. One or the major efforts of the actors in this process is to reformulate arguments in a way that makes them suitable for the pregiven social constellation and interest structure.

The transformation of one's own arguments into the language of another co-actor serves the following purposes:

(i) it imports credibility, and thus legitimation, by the appeal to external arguments that are used as unassailable resources;
(ii) it enables the carrying out of (internal) innovation processes

because they can be defined in respect to the outside as follow-up actions;

(iii) it helps to relabel planned actions and programs, thus immunizing them against critics and other risks; and

(iv) it makes the co-actor jointly responsible for the maintenance of one's own position and forces him to act.

Thus the transformation of arguments helps to bring formerly inconsistent positions together — such as military and unemployed (because of the jobs the military industry can supply), or military and ecology (because of the capabilities of pollution detection by space satellites), and so on. But while producing a great deal of external support, the adjustment of interest entails also the necessity of political and technical compromise. It is no longer possible to pursue one's own interests in a straightforward manner, for the redefinition of problems in terms of the perceptions of the co-actor also means a partial departure from strong selfish interests. For example, the German defense industry would prefer national arms programs; but because these are not politically acceptable, the industry is forced into international cooperation despite such perceived hazards as, for example, the danger of "technological impoverishment" (Lamatsch 1986).

Compromises are often painful if compared with short-term interests. But in the long run, the way of compromise mostly serves to better all interests. Through negotiation the perceptions of other actors and of potential refusers come into play, and the outcome of the negotiation process is more stable than pigheadedly enforced politics. Because the result of successful negotiations is an emerging process of mutual reinforcement, the system then gains a strength and self-dynamics that a sole actor could never gain. And the resources of mutual legitimation are — as has been shown above — much more manifold in such a social structure than in a monolithic system. This may explain the dynamics and innovation capacities of such structures, as well as their stability and resistance to "ex"-external influences.

But is has to be asked how innovations (technical as well as social) of this kind can emerge, given that the different co-actors stem from fields far removed from each other. The construction of social networks can be explained as follows (53): Each social system (politics, economy, science, and others) is conservative, to the extent that the working of the internal mechanism does not produce any innovation or change. But the maintenance of a system entails a dependence on resources supplied by the environment, and the inclusion of the environment leads to disorder inside the system. (Some environmental facts may be results of the actions of other systems, which in that way rearrange the frame of reference for their co-actors.) Innovation can now take place if this disorder works as an amplifier for nonconservative, mostly marginal, positions. The balance of power within the field can be rearranged if such positions make use of the environment by importing arguments, or resources (e.g., research funding), or credibility (as in the case analyzed above). A special situation occurs if, in different fields, actors are in the play who have analogous or even common interests; this may lead to hybrid structures (as in the space field), which may even make themselves independent. These processes are intensified by the fact that each system is forced to speak to its environment in a non-systemic, but common, language. Special problems cannot be represented to the outside without this language. Such transformations to another language, or even another problem's perception, in turn shape the self-image of the system (as, for example, the notion of the social usefulness of science has increasingly become part of its image, even if it is not necessary for the functioning of its internal logic).

When an innovation has been achieved and a network has been constructed, in a second stage the new social structure begins to stabilize itself by processes of autonomization and immunization. The network increasingly becomes the frame of reference, though the actors are still part of the field they depend on. But attempts to extend the borders of the hybrid structure (even if not a social system in a strict sense), and to expand the area of influence, show a distinct shift from the construction of internal coherence to the struggle with other external claims. The mid-eighties may be regarded as the transition from the first to the second stage in space policy in Germany. Manned space flight has gained a self-dynamics that begins to exceed even the genuine interests of the different actors and thus seems to be unstoppable. The build up of military space technology (which many of the actors surely do not ask for) appears more and more inevitable in order to avoid risking jobs, security, the future, and so on.

We can only speculate about the consequences for science (and for R&D policy as well) if the involvement into structures (such as the space lobby) grows, or, more generally, if network structures, affecting science in a similar way, become tighter. The externalization of the definition of problems (54) and the partial dissolution of the internal logic of science are, in themselves, not the crucial problem, as the model "finalization of science" has already pointed out (see Böhme, Daele, and Krohn 1973). If science generally is open for external control, then militarization is one of the results possible. But it is in no way a constraint, for other alternatives exist. And the model used in the present analysis shows a mechanism that may just as easily lead to the current situation in space policy as it may lead out of it again. The public has been confirmed as a crucial factor, together with the peace movement, and scientific and political organizations. All these elements are part of the process, even if they sometimes feel helpless in view of the power of the military-industrial complex. Every actor in the field actively shapes the frame of reference inside which the other actors can only move. The case of manned space flight in Germany is proof of the fact that the emergence and development of technology can be reconstructed as a line of selection of alternatives that are generated and shaped by social processes. Science is both part and object of the game, the outcome of which depends on the force of each actor to shift the balance of power.

Acknowledgments

Many thanks to Peter Weingart, who discussed former versions of this paper and gave important impulses that helped to write the present version. Wolfgang Krohn and Rainer Rilling also gave important advice.

Notes

1. Because several defense and space projects of the 1970s, as well as the contracts for building nuclear-energy plants, will run out during the late 1980s, this complex

calls for follow-up orders visionary enough to allocate resources for at least ten or fifteen years (comparable to U.S. Star Wars). The chairman of the German Fraunhofer-Gesellschaft, Max Syrbe, referred to this point when he said: "The research capacities that became idle due to the decline of nuclear reactor technology, are already budgeted for these projects [Hermes, Columbus, etc.]" (*Bild der Wissenschaft*, May 1986, p. 79). And the president of the German aerospace research organization (DFVLR), Hermann L. Jordan, proposed a simple measure: "redirection in the personnel sector: from nuclear technology to space flight" (Stellungnahmen, p. 9). (All German-English translations are by J. W.).

2. Recent news confirms this: Foreign Minister Genscher has made himself a spokesman for German participation in manned space flight and the military use of space technologies (*Frankfurter Allgemeine Zeitung* [hereafter, *FAZ*], 28 March 1987). The government increasingly seems to be adopting the plans of the space lobby, even though the terminology used in public is slightly more moderate.
3. See Brauch 1984:34, and Weyer 1985.
4. See *Bild der Wissenschaft*, 1968, p. 836.
5. See *Bild der Wissenschaft*, 1971, pp. 691—697.
6. See *Bild der Wissenschaft*, 1972, pp. 809—817.
7. See, e.g., Anhörung, and Feuerbacher 1984:52.
8. For details, see section IV, below.
9. For details, see section III, below.
10. See also Daele, Krohn, and Weingart 1979b, and Hoch 1988. The concept "hybrid structures" was first used by Daele et al. (p. 27) in the context of R&D policy. Krohn further developed this idea and added the concept of a feedback mechanism, thus going beyond classical concepts of unidirectional steering (see Krohn 1981).
11. Even if stimulated by military interests; see Wilkes 1978:5.
12. *Der Spiegel*, February 1985, p. 79.
13. See *Der Spiegel*, March 1984, pp. 61—62, and February 1985, p. 77.
14. *Bild der Wissenschaft*, 1984, p. 54.
15. See above.
16. Which can hardly be justified, because the technologies used in space flight are usually those of yesterday; see Keppler 1986:536, and Köhler 1984:51.
17. *Wehrtechnik*, March 1987, p. 16.
18. See below.
19. *Bild der Wissenschaft*, May 1986, p. 145.
20. See the scenarios in IABG 1986.
21. *Wehrtechnik*, July 1986, p. 39.
22. *Ibid.*, p. 40.
23. See Anhörung. Here again, the written opinion of the MBB representative Othmar Heise shows a more realistic viewpoint when he admits that positive effects of space research on economics cannot be proved (Stellungnahmen, p. 13).
24. The first common NATO satellite was launched in 1970, nearly coincident with the first German civilian satellite; see Büdeler 1979:369.
25. See below.
26. DFVLR 1986, pp. 5/2, 6/2—4; see also *Wehrtechnik*, March 1987, p. 65.
27. See Schreiber 1986. This article was published in *Europa Archiv*, which can be regarded as a semiofficial organ of the federal government. A short sample:

"Thereby space research is of the highest importance in terms of security and power policy, because it corresponds to the conquest of the seas and of airspace in the past" (pp. 636–637).

28. See also Dickson 1985:1244.
29. This shift can easily be shown by looking at the technical outfit of the Ariane family:

Model	*Payload*	*Orbit*	*Comments*
Ariane I	1,780 kg	GTO	production ends 1988/99
Ariane II	2,177 kg	GTO	further devel. of A I
Ariane III	2,580 kg	GTO	further devel. of A I
Ariane IV	ca. 4,500 kg	GTO	further devel. of A III; 1st flight planed 1986/87
Ariane V	5,200 kg	GTO	new devel., for the 1990s
	or 15,000 kg	LEO	

(Data from *Bild der Wissenschaft*, February 1986, p. 50; *FAZ*, 28 May 1986.)

If we assume (a point I cannot discuss here at length) that the geo-orbit is primarily interesting for commercial users, while the low orbit can contribute very little to commercial, scientific, and probably economic concerns, and if we further take into consideration that the heaviest communications satellites ever built weighed about 2,000 kg, while military satellites weigh 12,000 kg or more, we can easily see that the new technological system the Europeans have decided to construct, Ariane V, has only little use for actual commercial or scientific purposes, but much more for military interests and/or for manned space flight by carrying the shuttle Hermes or elements of a space station into low orbit.

30. This example of European space projects also shows the enormous self-dynamics that a program of this kind can gain. When the German government decided in January 1985 to participate in the Ariane V program and the Columbus program, but refused at the same time to do any other projects, everyone knew — and the space lobby used this argument with vehemence during the parliamentary hearing (see Anhörung, pp. 111, 165) — that the space program only makes sense if it is completed by the missing link: a shuttle, launched by Ariane to transport materials and astronauts to Columbus. So predecisions were made that produced irreversibilities and constraints to continue on the preestablished path.
31. A new launching plant, a set of satellites, and other details had to be added; see DFVLR 1986, and Johanson 1987.
32. *FAZ*, 25 August 1986; DGAP 1986:55–56.
33. Fricke, in Stellungnahmen, p. 11.
34. Here even ignoring the German plans for a futuristic shuttle called Saenger, to be developed during the construction period of Hermes.
35. See *FAZ*, 21 June 1986.
36. As the *FAZ* (20 June 1986) does it.
37. The case of the discussion about a military version of the German Airbus-plane, carefully launched in 1986, shows the way this mechanism works, or starts to

work, especially in a situation of economic crisis and unemployment.
38. *Bild der Wissenschaft*, May 1986, p. 146.
39. Heise, in Stellungnahmen, pp. 8, 14.
40. Heise, *ibid.*, p. 19; Hartbaum, *ibid.*, p. 6.
41. See, e.g., Riesenhuber, in "Das Parlament," p. 3.
42. *FAZ*, 17 December 1986.
43. Schmidt, in Stellungnahmen, p. 3.
44. Heise, *ibid.*, pp. 17—18; for more details, see Fuchs 1986.
45. "Das Parlament," p. 12.
46. *Ibid.*
47. The notion "shelterbelt" has been borrowed from Lakatos (1974) — not in a strict interpretation, but in an analog transfer.
48. If we look at the history of space flight, this suspicion is very quickly confirmed; see Brauch 1984, and Büdeler 1968.
49. See *FAZ*, 4 August 1986.
50. In *Spektrum der Wissenschaft*, March 1986.
51. This sector is even so profitable that government can, after having paid enormous costs for R&D, withdraw from this field, leaving the profit-collecting to the industries.
52. See Anhörung, and *FAZ*, 4 August 1986.
53. I refer to the recent research of Wolfgang Krohn and Günter Küppers; see Krohn 1985, Krohn 1986, Krohn and Küppers 1987, and Küppers and Paslack 1986.
54. In R&D policy a similar externalization can be discovered, if we look at the step-by-step replacement of the definitions of social problems (such as health, work, ecology) by external problems (such as world market, foreign policy, conquering the deep seas or outer space).

Bibliography

Abstein, Günter. 1986. "Zivile Nutzung von Rüstungsprojekten." Paper delivered at conference on "Economic and Technological Aspects of Armament," 8—10 December 1986, Wildbad-Kreuth.

Althanz, Peter, et al. 1984. *Militarisierung des Weltraums.* (Schriftenreihe Wissenschaft und Frieden, vol. 2). Marburg: Bund demokratischer Wissenschaftlerinnen und Wissenschaftler.

Anhörung: Deutscher Bundestag. 1985. "Weltraumforschung — Weltraumtechnik." *Ausschuss für Forschung und Technologie, Stenographisches Protokoll der 46. Sitzung des Ausschusses, 11—12 November 1985*, öffentliche Anhörung. (Mimeographed manuscript.)

BDLI (Bundesverband der Deutschen Luftfahrt- und Ausrüstungsindustrie, e.V.). 1984. *BDLI-Memorandum zur Zukunft der Raumfahrt in der Bundesrepublik Deutschland.* Bonn.

BMVg (Der Bundesminister der Verteidigung). 1986. *SDI: Fakten und Bewertungen, Fragen und Antworten, Dokumentation.* Bonn.

Böhme, Gernot; van den Daele, Wolfgang; and Krohn, Wolfgang. 1973. "Die Finalisierung der Wissenschaft." *Zeitschrift für Soziologie* **2**:128—144.

Brauch, H. G. 1984. *Angriff aus dem All: Der Rüstungswettlauf um Weltraum*. Bonn: J. H. W. Dietz.

Büdeler, Werner. 1968. *Aufbruch in den Weltraum*. Munich: Ehrenwirth.

—— 1978. *Raumfahrt in Deutschland: Forschung, Entwicklung, Ziele*. Frankfurt/ Berlin/Vienna: Econ Verlag.

—— 1979. *Geschichte der Raumfahrt*. Künzelsau/Talwil/Strassburg/Salzburg: Singloch.

—— 1982. "Transportsysteme bis ins All: Forschung und Entwicklung in der deutschen Industrie. Zum Beispiel: MBB." *Bild der Wissenschaft*, 1982, pp. 64—79.

Bundeshaushaltsplan für das Haushaltsjahr. Bundestagsdruckasche. Bonn.

Daele, Wolfgang van den; Krohn, Wolfgang; and Weingart, Peter. 1979a. *Geplante Forschung: Vergleichende Studien über den Einfluss politischer Programme auf die Wissenschaftsentwicklung*. Frankfurt am Main: Suhrkamp.

—— 1979b. "Die politische Steuerung der wissenschaftlichen Entwicklung." in Daele, Krohn, and Weingart 1979a:11—63.

DFVLR (Deutsche Forschungs- und Versuchsanstalt für Luft- und Raumfahrt). 1984. *Strategiestudie Raumfahrt*. Cologne.

—— 1985. *Jahresbericht 1985*. Cologne.

—— 1986. *Programmbudget 1986*. Cologne.

DGAP (Forschungsinstitut der Deutschen Gesellschaft für Auswärtige Politik). 1986. "Deutsche Weltraumpolitik an der Jahrhundertschwelle: Analyse und Vorschläge für die Zukunft." Experts' Group Report (Chairman: Karl Kaiser). Bonn.

Dickson, David. 1985. "A European Defense Initiative: The Idea That European Nations Band Together for a Strictly European Version of SDI is Gaining Support." *Science*, 20 September 1985, 1243—47.

"Faktenbericht 1986 zum Bundesbericht Forschung." Bundestagsdrucksache 10/5298. Bonn.

Feuerbacher, Berndt. 1984. "Die neue Raumstation: Lohnt sich der Flug ins All?" *Bild der Wissenschaft*, 1984, pp. 48—55.

Fuchs, Katrin. 1986. "'Europäische Verteidigungsiniative': Einstieg in SDI." In *Sozialdemokraten fordern: Aus Friedenssehnsucht praktische Friedenspolitik machen* ed. Rolf Seeliger, pp. 82—91. Munich:

Gambke, G.; Kerscher, R.; and Kertz, W. 1961. "Denkschrift zur Lage der Weltraumforschung." Commissioned by the Deutsche Forschungsgemeinschaft. Wiesbaden.

Haunschild, Hans-Hilger. 1986. "Zivile Forschungsförderung und Perspektiven der Technologieentwicklung." *Wehrtechnik*, July 1986, pp. 61—64.

Hoch, Paul K. 1988. "The Crystallization of a Strategic Alliance: The American Physics Elite and the Military in the 1940s." In this volume.

Högenauer, Ernst. 1986. "Mit Hyperschall ins All: Europas Beitrag zur Eroberung des Weltraums." In "Das Parlament," p. 12.

IABG (Industrieanlagen-Betriebsgesellschaft mbH). 1986. *Entscheidungsstrukturen und Entscheidungsprozesse im Raumfahrtbereich der Bundestrpublik Deutschland*. Ottobrunn.

Johanson, Anatol. 1987. "Bemannte Raumfahrt: Zweifel an Europas Träumen." *Bild der Wissenschaft*, January 1987, pp. 38—44.

Junker, Reinhard. 1986. "Forschung in der Bundesrepublik — Entwicklung, Strukturen und politische Steuerung." Paper delivered to science policy conference: "Zwischen Hoffnung und Bedrohung," 5—7 December 1986, Cologne.

Keck, Otto. 1984. *Der Schnelle Brüter: Eine Fallstudie über Entscheidungsprozesse in der Grosstechnik.* Frankfurt/New York: Campus Verlag.

Keppler, Erhard. 1986. "Raumfahrttechnik 1986 — Entwicklungsstand und Perspektiven." *Gewerkschaftliche Monatshefte* **37**:532—541.

Kitschelt, Herbert. 1979. *Kernenergiepolitik: Arena eines gesellschaftlichen Konflikts.* Frankfurt/New York: Campus Verlag.

Köhler, Horst W. 1976. "Neue Werkstoffe aus dem Skylab-Schelzofen." *Bild der Wissenschaft,* 1976, pp. 78—84.

—— 1984. "Amerikas nächster Schritt im Weltraum — die Raumstation." *Bild der Wissenschaft,* 1984, pp. 50—51.

Krohn, Wolfgang. 1981. "Thesen zur politischen Steuerung der Entwicklung von Wissenschaft und Technologie." In J. v. Kruedener and K. v. Schubert, *Technikfolgen und sozialer Wandel,* pp. 167—176. Cologne: Verlag Wissenschaft und Politik.

—— 1985. "Wie konstruktiv ist der Konstruktivismus? Eine epistemologische Auseinandersetzung." Conference at the meeting of the DGS section, "Science Studies", 16 November 1985. (Unpublished paper.)

—— 1986. "Zur Entwicklung der Wissenschaft." (Unpublished paper.)

Krohn, Wolfgang, and Küppers, Günter. 1987. "Wissenschaft als selbstorganisierendes System." Bielefeld. (Discussion paper.)

Küppers, Günter, and Paslack, Rainer. 1986. "Theoriendynamik und Weltbildwandel am Beispiel der Selbstorganisation." (Unpublished paper.)

Lakatos, Imre. 1974. "Falsifikation und die Methodologie wissenschaftlicher Forschungsprogramme." In *Kritik und Erkenntnisfortschrift,* ed. Imre Lakatos and Alan Musgrave, pp. 89—189. Braunschweig: Vieweg, Fried., & Sohn Verlagsgesellschaft.

Lamatsch, Hagen. 1986. "Die Bedeutung der Rüstungsindustrie für die Bundesrepublik Deutschland." Paper presented at the conference: "Economic and Technological Aspects of Armament," 8—10 December 1986, Wildbad-Kreuth.

"Das Parlament: Die Woche im Bundeshaus." 1986. Thematic issue of *EG-Forschungspolitik,* nos. 33—34.

Radkau, Joachim. 1983. *Aufstieg und Krise der deutschen Atomwirtschaft 1945—1975: Verdrängte Alternativen in der Kerntechnik und der Ursprung der nuklearen Kontroverse.* Reinbek bei Hamburg: Rowohlt.

—— 1986. "Angstabwehr: Auch eine Geschichte der Atomtechnik." *Kursbuch* **85**:27—53.

Scheffran, Jürgen. 1984. "Die Europäische Weltraumgemeinschaft — Aufbruch in die Zukunft?" *Informationsdienst Wissenschaft und Frieden* **5**:8—10.

—— 1985. "Die Europäische Weltraumgemeinschaft — Aufbruch in die Zukunft?" *Blätter für deutsche und internationale Politik,* 1985, pp. 169—185.

Schreiber, Wolfgang. 1986. "Die Bedeutung der Erforschung und Nutzung des Weltraums für die militärische Sicherheit." *Europa Archiv* **21**:629—638.

Stamm, Thomas, 1981. *Zwischen Staat und Selbstverwaltung: Die deutsche Forschung im Wiederaufbau 1945—1965.* Cologne: Wissenschaft und Politik.

Steinecke, Wolfgang. 1986. "Möglichkeiten und Grenzen internationaler Rüstungskooperation." Paper delivered at the conference: "Economic and Technological Aspects of Armament," 8—10 December 1986, Wildbad-Kreuth.

Stellungnahmen der Experten. 1985. Submitted in preparation for the 46th meeting of the Committee on Research and Technology of the German Bundestag on November 11—12, 1985, in Bonn. (Unpublished papers.)

Van Allen, James A. 1986. "Bemannte Raumstation: Schaden für die Forschung?" *Spektrum der Wissenschaft*, March 1986, pp. 36—44.

Weltraumforschung. "Nutzen für Alle." An exhibit by the Federal Minister of Research and Technology. (n.p., n.d.)

WEU (Western European Union, Assembly of). 1984. Thirtieth ordinary section, "The Military use of Space." Report submitted on behalf of the Committee on Scientific, Technological, and Aerospace Questions by Mr. Wilkenson, Rapporteur. Doc. 976 (15 May 1984) (First part); Doc. 993 (8 November 1984) (Second part).

Weyer, Johannes. 1985. "Chaos oder System? Anmerkungen zur Wissenschaftspolitik der Faschismus." *Forum Wissenschaft*, no. 2, 1985, pp. 31—35.

—— 1986. "Die Bundesrepublik am Scheidewege: Forschungspolitische Alternativen im Kontext der SDI-Diskussion." Paper delivered at the forum, "Science for Peace," 14 July 1986, Bielefeld. (Unpublished paper.)

Wilkes, Owen. 1978. "Military Research and Development Programs: Problems of Control." *Bulletin of Peace Proposals* **8**:3—10.

MIX
Papier aus verantwortungsvollen Quellen
Paper from responsible sources
FSC® C105338

If you have any concerns about our products,
you can contact us on
ProductSafety@springernature.com

In case Publisher is established outside the EU,
the EU authorized representative is:
Springer Nature Customer Service Center GmbH
Europaplatz 3, 69115 Heidelberg, Germany

Printed by Libri Plureos GmbH
in Hamburg, Germany